Survive and Thrive
A Guide for Untenured Faculty

Survive and Thrive: A Guide for Untenured Faculty

Wendy C. Crone

www.morganclaypool.com

ISBN: 9781608455133   paperback
ISBN: 9781608455140   ebook

DOI 10.2200/S00234ED1V01Y201009ENG011

A Publication in the Morgan & Claypool Publishers series
Lecture #11
Series ISSN
Print 1939-5221   Electronic 1939-523X

Survive and Thrive

A Guide for Untenured Faculty

Wendy C. Crone
University of Wisconsin–Madison

MORGAN & CLAYPOOL PUBLISHERS

ABSTRACT

The experience of an untenured faculty member is highly dependent on the quality of the mentoring they receive. This mentoring may come from a number of different sources, and the concept of developing a constellation of mentors is highly recommended, but a mentoring relationship that is guided by the mentee's needs will be the most productive. Often, however, the mentee does not know their own needs, what questions to ask, and what topics they should discuss with a mentor. This book provides a guide to the mentoring process for untenured faculty. Perspectives are provided and questions posed on topics ranging from establishing scholarly expertise and developing professional networks to personal health and balancing responsibilities. The questions posed are not intended for the mentee to answer in isolation, rather a junior faculty member should approach these questions throughout their untenured years with the help of their mentors. *Survive and Thrive: A Guide for Untenured Faculty* will help to facilitate the mentoring process and lead junior faculty to a path where they can move beyond just surviving and truly thrive in their position.

KEYWORDS

mentoring, faculty, tenure, career planning

To my family.

Contents

Preface

INTRODUCTION FOR THE UNTENURED FACULTY MEMBER

As you work toward tenure, keep three basic tenets in mind. First, you were selected from a large and competitive group of applicants. Second, your colleagues at your institution have an interest and investment in your success; it is important that you foster this interest. Third, you need to take a professional interest in your development as a faculty member.

This book is intended to facilitate a proactive and rewarding approach to your professional development and faculty experience. Actively pursuing your goals will increase your professional opportunities and successes, allowing you to thrive in your new faculty position. You should customize the issues and questions in this book to your particular discipline, institution, and personal goals.

Questions are posed for your consideration through this book. Some questions are appropriate at different stages in your career as a faculty member. It is not intended that you read this book from cover to cover in one sitting. Attempting to address too many questions all at once can easily make you feel overwhelmed. Similarly, imagining that one could take on every question in the book simultaneously is unreasonable. Also, it is important to keep in mind that not every question will apply to you. Use this book as a resource recognizing that no single prescription for success in a faculty position is available. The differences in the practices of your particular field or the emphasis of your particular institution will determine which questions are relevant to you. Also, keep in mind that you are a unique individual with your own unique professional and personal circumstances, thus you must determine those questions that apply to you and how to prioritize their importance.

Addressing the questions should be a process. You are not supposed to know the answers a priori. You will be most successful if you use this book in conjunction with your mentors. If you have a formal mentoring relationship with a senior faculty member, you may want to address a section a month during your regular lunch or coffee meetings over the course of the year. You might use this book to help stimulate and direct your conversations. If you have multiple formal or informal mentors you can go to, this book can help you to prompt conversations about certain aspects of your career which will be helpful to you. Successful faculty often have a "constellation" of mentors that they seek advice and guidance from throughout their career.

Although you have the biggest stake in your success, your colleagues and your institution also have an interest in seeing you succeed. You will find that there are many people on your campus and within your discipline willing to assist you in your professional development. You will need these people, so seek them out and nurture these relationships; colleagues and mentors are critical to your career success. This book will help you to be better prepared for discussions with your mentors and colleagues. After all, the right answers only come when you to ask the right questions.

In many cases, the questions posed will be associated with specific actions you will need to take. Because you cannot act on every front simultaneously, you will need to set a wide range of goals for yourself - some you know you will accomplish quickly and some that may take more time. Along the way keep good records of your progress. Documentation of some of the information discussed in the book will be important for your tenure packet. Take time to recognize and reflect on your accomplishments regularly. Share your achievements with your department head, review committee, and mentors. And, finally, remember to celebrate your successes!

This book is organized according to six key components of your professional development:

- **Tough Questions About Why You Are Here**

- **Joining Your Department and Discipline**

- **Establishing Expertise**

- **Developing Networks, Relationships, and Mentoring Activities**

- **Getting Support and Evaluating Your Personal Health**

- **Planning for the Future**

You may not want to consider these issues sequentially, so an index is also provided as the end to facilitate consultation on a specific question or topic. Because your professional development is your responsibility, you must make the time to develop and act on a strategic plan that addresses your specific needs and goals. Take ownership of your career!

This guide was developed to help you get your career off to a good start and keep it on track.

Don't just survive at your university - thrive!

INTRODUCTION FOR THE MENTOR

This book is also meant for mentors. I have often interacted with senior faculty who hope to be a good mentor to a junior colleague but are not sure what advice or topics of conversation might be helpful. This book can provide a mechanism for identifying questions that need to be discussed or structuring a series of mentoring interactions. This book is organized according to six key components of professional development:

- **Tough Questions About Why You Are Here**

- **Joining Your Department and Discipline**

- **Establishing Expertise**

- **Developing Networks, Relationships, and Mentoring Activities**

- **Getting Support and Evaluating Your Personal Health**

- **Planning for the Future**

Some people have successfully implemented the book in their mentoring relationship by addressing each section at a monthly meeting, while others have used the book to prompt discussion on topic areas of immediate interest to the mentee. An index is provided at the end of the book to facilitate consultation on a specific topic.

Because not every question will apply to your mentee, part of your job will be to help them decide which ones are important to focus on and at which stage in their career development they should be addressed. Use this book as a resource, recognizing that differences in the practices of a particular field or the emphasis of a particular institution will determine which questions are relevant. Your experience is the key ingredient to making the mentoring relationship work.

Mentoring relationships come is various forms and have different levels of formality and expectations. The term "mentor" also means different things to different people. To some, it connotes teacher, advisor, and counselor, while to others, there is a either a deep friendship implied or a substantial power relationship at play. Mentees may be looking for different things from their mentor. For instance, you may be expected to provide positive and constructive feedback, understanding and empathy, encouragement and nurturing, and assistance in developing networks. Before embarking on a new mentoring relationship, however, you need to ask yourself which roles do you feel comfortable fulfilling and what time commitment you can give to the mentoring relationship. There may also be an added complexity to the relationship because you may also have to fulfill an evaluation role with a particular mentee. This potential source of conflict should be addressed up front.

Unless it is a brief and fleeting interaction, you should make a point to discuss the parameters of the mentoring relationship with your mentee. There are a number of questions you should consider: How much time can you each devote? How frequently should you plan to meet? Will you do all of your mentoring in person, or are phone and email exchanges also useful? On which topics do you feel comfortable mentoring that fall within the needs of your mentee? Can you develop a productive and non-threatening relationship? How does the mentee best take criticism and constructive feedback? Will it be beneficial for you to create a mentoring plan that you will enact over time?

Wendy C. Crone
August 2010

I recall wanting to be a professor as early as grade school after visiting the small college where the mother of one of my friends was on the faculty. Early in my graduate school career, I became disillusioned with this goal as I saw more of the day to day realities of faculty life. After some experience with a variety of academic institutions, however, I learned that faculty positions differ from institution to institution and even from position to position within the same department. This was a wonderful realization. It helped me to rekindle my old dream, and it helped me to identify the type of institution I would be happy in. Because of this and later experiences, I believe that institutional fit is a critical component to happiness in a faculty position.

CHAPTER 1

Tough Questions About Why You Are Here

Occasionally, in life, one stops, looks around, and wonders - how did I get here? Many faculty chose an academic path early in life, but there are also some who come upon academic life later or by accident. Most assistant professors fall into one of two categories: those who are already committed to this career path or those who are still exploring a variety of opportunities. Regardless of which category into which one falls, it is important to stop, look around, and engage in self-assessment regularly.

Don't wait until you face an obstacle or setback; a time of crisis may not be the best point to question the path you have chosen. However, careful planning may allow you to smooth the path and avoid the obstacles in the first place. Planning and self-assessment are important ingredients for a successful career.

With any career choice there are a number of things one must compromise in the rest of our lives in order to do the job successfully. Don't continue blindly without reflection, or just because you had already set yourself on this path. The job of an assistant professor in today's academic environment is challenging! There are difficulties your more senior colleagues may not have had to face. So, at least once in the early stages of your adventure, stop, look around, and ask yourself - why am I here? Your answer to this question may reenergize you and recommit you to the challenge you have taken up or it may prompt you to consider a different path.

This section breaks down the very broad question of, "Why am I here?" into more manageable bites. These are not all the possible questions, simply a set to help get you started. Strategy we will pursue throughout this book begins with overarching questions to consider, followed by mentoring conversation essays to reflect on, and then detailed questions and suggestions to address by yourself and with you mentors. Do not expect to have all the answers to the questions being posed. The big answers to the big questions may only come with time, and reflection, and through discussion with your trusted colleagues and advisors.

Read and reflect on the questions below individually or as with one of your mentors. Make some notes. Identify more questions of your own. Select some for discussion with your mentor.

1.1   ASSESSING THE FIT

1.1.1   OVERARCHING QUESTIONS TO CONSIDER

- What career paths do you envision for your future?

- Is your current institution the place you would like to spend your career?

- What is your back up plan if tenure is not in the cards?

1.1.2   MENTORING CONVERSATION: ON INSTITUTIONAL FIT

An academic career may not be the right choice for everyone, but I would argue that within academia there is such a wide range of options so that one can often find a good fit. I frequently have conversations with graduate students and post docs about what might be the right choice for them. In some cases, a student will bluntly say, "I don't want your kind of job." For a variety of reasons they have decided that what they see me spending my time on as a faculty member at a big research university is not what they want to do with their life. (On some days, I might agree with them myself.)

Although I don't try to push every student towards an academic career, I do want them to realize that not every job in academia looks like mine. Not only are there a number of different types of institutions (Doctorate-granting Universities, Master's Colleges and Universities, Baccalaureate Colleges, Associate's Colleges, Special Focus Institutions, and Tribal Colleges)[1], they each have different personalities (as a result of their public vs. private nature, or religious affiliation for instance), and there are different positions both on and off the tenure track. Depending on a person's interests and long term goals, they may be more interested in an instructor or laboratory coordinator position than in a tenure track faculty position. At each institution, the requirements of even the tenure track faculty positions can be quite different, ranging from only teaching to heavily focused on research.

Other students come to me with a keen interest in obtaining an academic position. I try to help them first identify what type of position might be best for them. The next step is to identify people in such positions who they can talk with to check their preconceptions with the realities of the position so that they can make sure the fit is right. I encourage them to find out about what it is like in that position and what qualifications search committees at these institutions expect to see. This often helps a student to identify holes in their background that they can get advice on how to fill, by identifying opportunities to pursue while conducting their current studies or what next position might serve as a stepping stone.

I have similar advice for faculty who find themselves in an academic position or particular institution that is not a good fit for them. In contrast to previous eras, it is not uncommon for faculty to move around. One's current position does not have to be a lifelong commitment. The key is to look for the type(s) of positions or institutions that would be a better fit and then develop the expertise

---

[1] "The Carnegie Classification of Institutions of Higher Education," The Carnegie Foundation for the Advancement of Teaching, `http://classifications.carnegiefoundation.org/index.php` (Accessed 11/11/09).

and record that would make you an ideal candidate for the position you want. This may involve shifting the emphasis of your current work (where possible), seeking out external experiences that would be valued, and building a network within your new target zone.

One caution however, when you are fact-finding about or even go so far as to interview for a new position, be careful not to come across too negative about your current position. It can give a generally negative impression of you. Instead, you can say "It's not a good fit, but…" go on to talk about some aspect you do appreciate or have excelled at in your current position. Then you can focus on why you feel the new position would be a good fit and your experience and achievements that are relevant to the new position.

1.1.3   DETAILED QUESTIONS AND SUGGESTIONS TO CONSIDER

1. What initially attracted you to your field of study? What continues to attract you?

2. In what ways are you satisfied with your progress in personal and professional development to this point? In what ways are you dissatisfied with your progress in personal and professional development to date?

3. Have you systematically examined the pros and cons of continuing on the academic path? Is it still in your best interest to continue on your current career path or would a different type of institution be more suitable? Consider:

   • Do you still have a passion for your field of study?

   • Are you/will you receive sufficient recognition for your work?

   • Have you evaluated your earning potential?

   • Are other more suitable positions available?

   • Have you considered how your personal and family responsibilities impact your career?

   • Do you have sufficient time for your other interests (hobbies, travel, volunteer work)?

4. What is the record in your department or institution in helping you and others in your position work towards obtaining tenure? Sometimes a hire is made without the intention of ever granting tenure. Can you determine if this is the case for you? Can you use the position to your advantage as a stepping stone to a better situation in which tenure is more feasible?

5. Have you explored the functions of the faculty at peer institutions? How does this compare to your institution? If there are discrepancies that are important to you, can you make change happen?

6. Feeling like you belong when you come from an underrepresented racial or ethnic group, different income or class background, sexual orientation, or gender in your field presents challenges. Will you choose to adopt strategies to "pass" or celebrate your "differences"?

7. The community in which we live is also an essential ingredient. Have you taken the time away from establishing your academic career to get to know the city you live in and the people around you? Are you living in an environment that you enjoy?

1.2   YOUR CAREER AND YOUR PARTNER

1.2.1   OVERARCHING QUESTIONS TO CONSIDER

- How do you and your partner's goals compliment each other?

- Have you had discussions about how you will make decisions affecting the both of you?

1.2.2   MENTORING CONVERSATION: ON NAVIGATING THE JOB HUNT WITH A PARTNER

Whether your partner has a career or not, he or she will have an influence on the decisions that play into taking a job offer and staying in the job you have. It is important for both of you to be comfortable with the professional opportunities available and the new community you will becoming a part of. I have known a number of candidates and colleagues who have declined offers or left institutions because the fit for the partner was not a good one. Reasons can range from the weather, to the community, to job opportunities. The point is, for a tranquil and supportive home life, both of you must be in agreement that this is the right place for both you.

Navigating the job hunt with a partner in the equation can certainly add variables, but it is wholly possible for you to find an agreeable solution. Many institutions recognize the "two body problem" and have mechanisms for spousal/partner job assistance both inside and outside of academia. Many smaller or more remotely located institutions have also realized that dual faculty hires can actually have advantages for the institution. Although I met my husband when we were both assistant professors at the same institution, we embarked on an external job hunt a few years ago. Even though there was only an open search in one of our fields, the institution we were considering interviewed us both and eventually made offers to us both.

From this experience, the most important lesson my husband and I learned was to communicate as much as possible, preferably before a decision is upon you. You need to have open conversations about your future goals, priorities, and preferences. Where you would like your career(s) to go? What hopes you have for your future life together? Also keep in mind that there is a good chance that one of you may end up making professional sacrifices for your common goals. You'll need to ask the question: Who is willing to make the sacrifice at this point in their career?

1.2.3   DETAILED QUESTIONS AND SUGGESTIONS TO CONSIDER

1. A number of creative strategies have been developed for work/life balance, especially for dual career couples. Can you explore the possibility of:

   - job placement assistance for your spouse/partner?

- dual career couple hires?
- maternity/paternity leave?
- stopping the tenure clock?
- job sharing?

2. Do you have a plan for when you will bring up the dual career issue in the job hunt process? After you have accepted an offer is too late. Depending on the circumstances, during the interview or the negotiation process will be most appropriate. Who can you consult with to determine the best timing for your situation?

3. Have you had an open discussion with your partner about:

- who's career takes priority in what circumstances and time frames?
- how will you negotiate job offers?
- how will family responsibilities be balanced with your career(s)?
- what is your back up plan if tenure is not obtained?

4. Are you and your partner in a agreement about the area of the country, size of city/town, and type of community you are comfortable living in? Have you considered the availability of day care and the quality of the school system?

5. At some point in your career, you may be approached by another institution or actively seek an outside offer. You may want to consider:

- Will your current department look at the offer in a positive or negative light?
- Will an outside offer hinder or help your tenure case?
- Are you really willing to leave your current institution?

BIBLIOGRAPHY
TOUGH QUESTIONS ABOUT WHY YOU ARE HERE

M.N. Bushey, D.E. Lycon, P.E. Videtich, **How To Get A Tenure-Track Position At A Predom-inantly Undergraduate Institution,** Council on Undergraduate Research, Washington, DC, 2001.

"The Carnegie Classification of Institutions of Higher Education," The Carnegie Foundation for the Advancement of Teaching, http://classifications.carnegiefoundation.org/index.php (Accessed 11/11/09).

A. B. Ginorio, **Warming the Climate for Women in Academic Science**, Association of American Colleges and Universities, Program on the Status and Education of Women, Washington, D.C., 1995

6 BIBLIOGRAPHY

"Jobs," The Chronicle of Higher Education, http://chronicle.com/jobs (Accessed 11/11/09).

J.M. Lang, **Life on the Tenure Track: Lessons from the First Year**, Johns Hopkins University Press, Baltimore, MD, 2005.

J. D. Spector, **Guide to Improving the Campus Climate for Women Commission on Women**, University of Minnesota, Minneapolis, MN, 1993.

When I first embarked on the endeavor of being a faculty member I discovered that, although I had spend a good many years of my life in the academy, the number of things I did not know about my academic appointment far outnumbered the things I did know. So I took this on as another research project, a puzzle to solve, a mystery to investigate. The other crucial bit of information I realized along the way was that, although answers to many questions were forthcoming, much of my research would be long-term. It takes time to acquire the needed information, determine the rules (written and unwritten), and find those key people who have the answers.

CHAPTER 2

Joining Your Department and Discipline

The first steps to achieving tenure can occur well before you take up residence at your new institution. Much of this part of the process will put you in fact finding mode - starting when you begin looking at ads for positions, and continuing through the untenured years.

To be successful in your new position, you will need the right tools. If you are very lucky, someone will tell you what these tools are and they will help you to obtain them. Unfortunately, most young faculty I have talked with did not have this experience. Often, it is not for lack of good intentions on the part of senior colleagues and mentors. The rules change over the years and the emphasis within the tenure committee may change as the committee members change. So the onus is on the junior folks. You must ask questions, search for information, and negotiate for what you need. Some of us feel more comfortable with some of these tasks than with others, but you must persist with them all!

Now that you have made the decision to join the academy in your particular discipline and profession, you will want to consider the following questions.

2.1 NEGOTIATING THE TERMS OF YOUR APPOINTMENT

2.1.1 OVERARCHING QUESTIONS TO CONSIDER

- Have you sought out advice or guidance that would help you enter into a successful negotiation?

- What aspects of your position and duties are negotiable?

- Can you do background research that will support your request?

2.1.2 MENTORING CONVERSATION: ON NEGOTIATING AN OFFER

Several years ago, one of the post docs in my group was offered an academic position. I encouraged her to negotiate the terms of the offer. This was not something she had intended to do. As a rule, I encourage everyone to negotiate, even if just a little, so that they can start to learn the art of negotiation and, hopefully, obtain the most optimal situation possible in the position they have been offered. I also believe the opportunity for negotiation can also help to set the tone in your new position, showing that you are a professional who knows what you need to be successful. The other point to remember is that you will very seldom have something given to you that you did not ask for. So, you must ask!

However, I should caution that you must ask for things that are reasonable, and you must ask in a professional and collegial way. There are a number of items that are negotiable, but two common topics are salary and teaching load. For both, a little research on the topic can go a long way. You can use your network to find out an amazing amount of information. This gives you information about the bounds and a strong foundation for negotiation. The post doc I mentioned earlier found out information about similar positions and had very good grounds for negotiation on several points. Although she was not able to directly negotiate her salary, the process was valuable because she found out that she was able to negotiate the amount of prior teaching experience counted towards her seniority, which ultimately set her pay rate. The process also helped her to decide if this was the right place for her.

Sometimes negotiation can get you more than what was initially offered, helping you to obtain things that will make you happier and more effective in your position. There are some choices you have concerning who to approach about negotiation. I usually suggest that one begin with the chair of the search committee or the chair of the department. Choose someone who can act as an advocate for you with the people who actually control the decisions (and the purse strings). The committee and/or department decided that you were the best person for the job, they want you to come to their institution, and they want you to succeed in the position.

2.1.3   DETAILED QUESTIONS AND SUGGESTIONS TO CONSIDER

1. What approach to negotiation will achieve the best outcome? How will your approach depend on the situation? Consider the strategies of approaching negotiation as:

   - one event in a long term relationship that you want to foster.

   - an opportunity for relationship building.

   - a collaborative undertaking.

   - an opportunity to promote and open. discussion that maximizes information flow in both directions.

   - a way to assess the needs of both parties.

2. Much negotiation of the terms of your appointment takes place before you accept an offer. The things that are negotiable depend on the type of institution and the department, but the primary concern should be to get what you need to enable you to be successful in the position. Items to consider at that time, or in the first year, include the following:

   - Start-up package (including money for your summer salary, graduate assistant's salary, postdoc funding, computers, equipment, conferences, and flexible funds for other costs)

   - Time limits on start-up package spending

- Salary
- Seniority granted for prior experience
- Moving expenses
- Teaching load (temporary reduction in teaching, semester off from teaching, choice of courses, control over when courses are taught)
- Office space and office furniture
- Laboratory, research or performance space renovated to your needs
- Computing facilities
- Job placement assistance for your spouse/partner

3. Your salary at the early stages of your career can have a dramatic impact on your lifetime earnings. Even a seemingly small dollar amount can grow to a large sum over the time frame of one's career. When approaching salary negotiation in an offer or at raise time:

  - develop a strategy in advance for the best approach to take with salary decision maker(s).
  - know what others in a similar field and at a similar level make.
  - set both a minimum and an upper goal.
  - don't undersell yourself in your opening negotiation.
  - don't concede to much too soon.
  - reiterate your points while remaining flexible.
  - conduct a mock negotiation with a friend to boost your confidence.

4. Have a frank discussions with your department chair about the following issues:

  - The track record of your department in supporting junior faculty
  - The availability of, and your eligibility for, financial support within the institution
  - Conditions you must meet for your appointment to continue
  - Teaching load and number of new preps each year
  - Courses you would prefer to teach
  - Release time for the development of new courses
  - Teaching assistantship support for the classes you teach
  - Teaching assistantships available for your graduate students
  - Expectations to buy out of a portion of your academic year salary
  - Vacation time and the amount of summer salary you are allowed to pay yourself
  - Preparation of your tenure packet

- Provisions for maternity leave, parental leave, medical leave, and elder care leave[1]
- Options for stopping the tenure clock for birth, adoption, elder care, or illness

5. If you already have or plan to have a family, it is important to find out about how your department and institution supports family responsibilities. In addition to reading up on the Family Medical Leave Act, you should also consider:

- obtaining a copy of your institution's maternity, paternity, and adoption policy.
- finding out about prior practice in your department and other departments in your college.
- talking to other faculty with a similar family situation to your own.
- discussing options for stopping the tenure clock with your chair.
- looking into how a change in family status will affect your benefits.

6. There is a long list of other items that you should ask about early on in the process. Some key issues in your field may include the following:

- Cost of a research assistant's salary and fringe benefits
- Percentage of overhead taken on your grants
- Funds available as matching money for grant proposals
- Number of graduate student applications coming into the program each year
- Quality of the graduate students in the program
- Office computer
- Computer networking infrastructure
- Support for technology enhanced learning
- Library services
- Shared facilities available for research
- Buyout policy
- Undergraduate advising load

7. There are a number of seemingly small issues revolving around departmental resource allocation that can affect how you are perceived in your department. Consider:

- What is viewed as a fair share of the office support for typing, photocopying, purchasing, etc.?
- Is there an established system for requesting library purchases?
- How will remodeling for your laboratory space be accomplished?

---

[1]J. C. Williams, "It's in Their Interest, Too," **The Chronicle of Higher Education**, August 31, 2006. http://chronicle.com/article/Its-in-Their-Interest-Too/46751/ (Accessed 11/11/09).

2.2 WHAT'S COMING?

2.2.1 OVERARCHING QUESTIONS TO CONSIDER

- Do you know what it takes to get tenure at this institution?

- What is your timeline to tenure?

- Are there options for coming up for tenure earlier or later?

2.2.2 MENTORING CONVERSATION: ON IT BEING MORE THAN JUST GETTING TENURE

There is a tendency for junior faculty to focus on and even obsess about tenure. Even though your purpose in taking a faculty position was not to guarantee yourself a job for life, it is easy to loose sight of your personal goals with the "ax" looming over your neck. Being denied tenure is not the end of the world (see Section 6.3) and, surprisingly, being granted tenure can feel anticlimactic.

Being aware of the requirements for tenure and working towards them are important, but don't loose sight of yourself in the process. The best strategy is to find an alignment between your own interests and the tenure requirements and pursue it with gusto. If your heart and mind are fully engaged, then you will perform at your best and achieve up to your potential. At many institutions, there is some flexibility to how scholarship is defined so that your tenure case does not have to look exactly like your colleague's.

A few years ago, two of my colleagues and I were asked to speak to a group of junior faculty at our institution about our tenure cases. It was a good panel because even though we are all three at the same institution and in similar fields, our cases looked quite different from even though we were all considered to be "success stories." We had each established ourselves in our respective research areas – a requirement for our institution – but we had struck very different balances between research, teaching, and service. Both the contrasts and similarities were helpful to see, and it became clear in our remarks that we had each been guided by our passions. Because we had focused on things we felt strongly about, we had more energy and enthusiasm for our work, and we were able to achieve more. Most institutions appreciate that they need a range of different kinds of faculty members to make the place work well and meet all of the institution's needs.

2.2.3 DETAILED QUESTIONS AND SUGGESTIONS TO CONSIDER

1. Have you taken time to get to know the institution you have joined? Look for information on the following topics:

   - The mission and vision of the university and college

   - Recent annual reports

   - Faculty policies and procedures

   - Tenure procedures and criteria

- Collective bargaining agreement (if faculty are unionized)
- Accreditation standards for the major and/or institution

2. What milestones have you set for yourself, and when do you plan to achieve them? What are the expectations of your department and institution? When does the first official performance evaluation occur?

3. Do you know what it takes to get tenure at your institution? What are the tenure metrics? How do you get the information you need about the requirements? Consider:

- Asking for written tenure guidelines
- Talking with your department chair, mentors, senior faculty inside and outside your department
- Talking with peers at your own and other institutions
- Taking advantage of orientations, workshops, and seminars designed for new faculty
- Attending professional conferences and meetings
- Observing the progress of others
- Observing the mistakes of others
- Reading general literature about academia and the tenure process

Remember that the rules change - what was true several years ago may no longer be the case!

4. When are the decision points for renewal of contract and/or tenure? Have you considered the tenure clock and how it fits into the rest of your life plans? In assessing the timeline, consider:

- departmental and institutional requirements/expectations.
- personal responsibilities (debt, child care, elder care responsibilities).
- possibility and desirability of stopping the tenure clock.

5. There are a number of larger factors that affect your job and how your energy is focused. Talk to people about the:

- vision of your department/college/institution.
- timeline of your faculty appointment.
- characteristics of the undergraduate and graduate student populations.
- methods used to recruit and retain students.
- facilities for teaching laboratory courses and equipment for using technology in the classroom.
- level of expectation for obtaining external funding.

- amount of secretarial and accounting support provided by the department.

- the methods of decision making used in the department.

- the characteristics of a successful faculty member in your department.

2.3   PROFESSIONAL ORGANIZATIONS

2.3.1   OVERARCHING QUESTIONS TO CONSIDER

- Are you becoming an active member of your professional organization(s)?

- Have you identified senior colleagues that you can connect with through professional organization(s) in your field?

2.3.2   MENTORING CONVERSATION: ON THE FINDING THE RIGHT PROFESSIONAL ORGANIZATION

Becoming a recognized member in your field can be done in a number of ways, but one of the best and most efficient mechanisms comes through joining a professional organization. In some fields, there is one choice – the organization that "everyone" belongs to – and in other fields, there may be multiple choices. Particularly if your interests are interdisciplinary, you may find that you need to make connections to more than one organization. There are also sometimes local or regional chapters of research and/or teaching-oriented societies that you might consider.

In my particular areas of research and teaching, there are far too many options. I could easily go to a dozen conferences a year, but I don't have the time, money, or inclination to do so. In the first few years of my faculty position, I went to the conferences of several different societies, some of them accompanying my senior colleagues. Each society and conference was different from the next, some in subtle and others in very distinct ways. I no longer regularly attend some of these conferences. In some cases, I found that my research was not a good fit, and in other cases, I found that the personality of the society was not a good fit.

At this point in my career, I have three societies that I am active in, regularly attending the conferences, giving talks, and participating in the society organization. One of these is a teaching-oriented professional society, and the other two span the interdisciplinary areas where my research lies. I have found the long term involvement to be fulfilling: allowing me to build relationships, identify collaborators, develop my professional reputation, and contribute to the future direction of my field.

One of these societies I have been a member of since my undergraduate days. My first conference participation was with this society in a student poster session. Over the years, this has grown into a connection that is almost like a second family. Not only are the other researchers friends who I enjoy seeing regularly, the society staff members are also wonderful people I enjoy interacting with year after year. At a recent conference I attended, I had lunches and dinners with several groups of friends, heard some great talks, connected with another researcher about a technique he developed that I am try on the material I work with, presented my own work, and got an invitation to write a journal paper for a special issue. So, sometimes these meetings can be fruitful in a wide variety of ways!

2.3.3   DETAILED QUESTIONS AND SUGGESTIONS TO CONSIDER

1. What are the important professional organizations in your field? Have you identified groups/organizations both inside and outside the university?

2. Can you afford the time and money to join all the "right" professional groups? Can you afford not to join them and attend meetings/conferences? What are your options if you do not have the financial resources to participate in these organizations? Will your department or college help support travel costs or membership fees?

3. Do you have a plan for which professional association conferences you should attend? If not, do you know where you can get this information?

4. Which scholars are highly regarded in your field? Are you familiar with their work? Can you create opportunities to interact with them at conferences?

2.4   JOURNAL ARTICLES, BOOKS AND OTHER SCHOLARLY PRODUCTS

2.4.1   OVERARCHING QUESTIONS TO CONSIDER

• What scholarly products are most highly regarded by your colleagues and your institution?

• Have you sought advice on the best places to publish or disseminate your work?

2.4.2   MENTORING CONVERSATION: ON WRITING

Every faculty position involves the production of intellectual products of some sort. Various contests and venues may be involved depending on your discipline, but written expression in some form is almost always required. There are a number of challenges we face when we approach writing. It seems especially difficult because some of our colleagues make it look so easy. I have found that writing has become easier with time, but I expect that even the most seasoned of us will at least occasionally face barriers.

The first common barrier is simply that of getting started. My personal strategy is to start on whatever part is easiest for me. As an experimentalist, it is quite easy to write the experimental techniques section of a journal article. If I am writing a research proposal, however, I usually start writing about the big picture first: what my grand goals are and how I think there will influence my field of research. For you, and depending on what it is that you plan to write, it may be different. You may also need to experiment to see what works best for you.

Writing can also be influenced by environmental factors. There are some things that I can write directly on a computer, while others demand pen and paper for the first draft. Changing the setting can also help me to overcome writing blocks. Leaving my office and walking over to a coffee shop with my computer can often get me started again. For other people, regiment is the most helpful thing. You may need to devote a specific time of day every day in the same location to get into a rhythm for your writing.

Advice on writing, particularly if you are approaching a format that is new to you, is also something you should consider seeking out. Senior colleagues as well as your peers can be valuable resources. They can provide guidance, samples of successful writing in the same format, and reading of your drafts. You could also consider joining or starting a writing group. This strategy can provide you with motivation for and feedback on your writing.

The worst thing you can do is avoid writing. Figure out a way, any way, to get yourself going. Also, remember that at some point you need to stop, show it to others, and eventually submit it. Recognize and come to terms with the fact that it will never be perfect. Even if you don't think it is quite ready yet, at some point you just have to be done.

2.4.3   MENTORING ACTIVITY: DEVELOPING A WRITING GROUP WITH YOUR COLLEAGUES

Step 1: Identify a few colleagues who are at a similar career stage and have common writing goals to your own. Have a conversation with each of them about their interest in forming a writing group. Identify the best venue for the meetings and timing for the meetings. Take into account that some members may have other obligations that prevent them from meeting at certain times or on certain days of the week.

Step 2: Develop consensus within the group about the formality of meetings, frequency of meetings, optimal size of the group, ideal group dynamics, and acceptable conduct for the group members. For instance, you might consider the following roles:

The author should:

- hold the position of authority in the conversation.
- begin by presenting their work and any specific questions about the writing.
- listen to the feedback provided.
- be gracious in the face of constructive criticism.

The group should:

- provide constructive feedback focusing on global issues rather than minutia.
- address the author's questions.
- include a complement in your comments by describing what is working well in the writing.

> • try to develop a dialog about the work while remembering to take turns and avoid interrupting.

Step 3: Set up an email group or listserve with the members of your writing group. As the "convener" of the group, you will be responsible for sending out reminders for meetings and keeping the momentum of the group going. It is good practice to rotate the "convener" responsibility to a new individual for a group that meets for more than one year.

Step 4: Set up the agenda for the next meeting at the end of the previous meeting. Ask each member to make commitments about what they will write prior to the next meeting, the deadline for sending around advance copies for review, and their willingness to review the work of the other group members prior to the meeting.

2.4.4   DETAILED QUESTIONS AND SUGGESTIONS TO CONSIDER

1. How is scholarship defined at your institution? In what types of scholarship[2] do you engage?

   - Discovery through original research
   - Integration and synthesis of knowledge
   - Application of knowledge in professional practice
   - Transformation of knowledge through teaching

2. Scholarly work requires disciplinary expertise and innovation, but it must also produce a product that can be evaluated. Be sure the scholarly endeavors that you engage in are approached with the following:

   - Clear goals
   - Thoughtful preparation
   - Appropriate methodology
   - Iterative improvement
   - Focus on outcomes

3. Contributions to the scholarly literature can come in many forms. Which forms are most appropriate for you and when should they be pursued?

   - Journal article
   - Conference proceeding
   - Technical report
   - Book or encyclopedia chapter

---

[2] R.M. Diamond, **Preparing for Promotion, Tenure, and Annual Review: A Faculty Guide**, Second Edition, Anker Publishing Company, Bolton, MA, p. 18-21, 2004.

- Edited collection or anthology
- Book on a scholarly research topic
- Performance
- Portfolio
- Workshop
- Textbook

4. Which journals are the most important and prestigious in your field? Do you read these regularly? Can you join or start a journal reading club?

5. In choosing where to publish your work, which journals are most prestigious for publishing your work? Which journals are most likely to publish your work? Is there a discrepancy between your answers to these questions? If so, how might you reconcile the differences?

6. Things you should consider when choosing a journal for your submission include the following:

- Appropriate topic
- Prestige
- Impact factor
- Readership
- Speed to publication
- Page charges
- Color page charges

7. In addition to writing the manuscript, submitting a journal article for publication involves several key steps. You should:

- have a colleague pre-review your work before submitting or present the work in a conference or seminar forum to obtain feedback.
- become familiar with the types of articles published by journals in your field to determine where your article fits best.
- find out the average time to publication.
- consider the prestige and acceptance rate of the refereed journals in your field.
- follow the guidelines and format of the specific journal to which you will submit the publication.

- include a letter to the editor or associate editor with the submission of your manuscript which contains the following information: the title, the journal for which you would like to have the paper considered, a brief summary of the manuscript and why you believe publication in this journal is warranted, a statement that the manuscript is not being considered for publication elsewhere, names of potential reviewers who could (or should not be) used to evaluate your work, and your contact information.

8. Generally, journal papers can only be submitted for consideration to one journal at a time, so it is important to track the progress of your submission. Often, it is possible to track it through the journal's website. When should you contact the editor if no response has been received? When should you withdraw your manuscript if a significantly long time period has passed?

9. A good paper should include the following:

   - Be on a topic worth investigating
   - Cite relevant literature
   - Have a clear hypothesis
   - Provide a logically constructed argument
   - Be clear and concise

10. Generally, a book prospectus can be simultaneously submitted for consideration to multiple publishers. What book publishers are most appropriate for you work? There are several types of publishers to which you might consider submitting your work:

    - University presses
    - Commercial scholarly publishers
    - University centers and institutes
    - Learned societies, museums, or libraries
    - Self-publishing or web-publishing

11. Some publishers have specific requirements for what should be included in the prospectus when submitting a manuscript for consideration. You should check the publisher website for author instructions. In general, a book prospectus will include:

    - a letter addressed to the acquisitions or executive editor summarizing the book, what is exciting about the project, and the intended audience,
    - table of contents,
    - sample chapter, and
    - your CV.

12. Evaluating feedback you receive on manuscripts is essential. Some feedback you receive will be helpful and constructive while other feedback may be unconstructive.

    - How will you decide whether to revise and resubmit or to try another journal or publisher?
    - How do you respond to reviewers' comments about your manuscripts?
    - How can you minimize the number of rewrites?
    - Do you have colleagues who can advise you?
    - Does your institution have a writing center or writing workshops that might be helpful?

13. If your manuscript is rejected, it is important to look at things objectively because rejection frequently occurs. Did you submit to the right journal or publisher? Have you asked a senior colleague to review and comment on your work?

14. As your book nears publication, work with your publisher to develop a marketing plan. Would it be appropriate to:

    - have your university communications office do a press release?
    - seek a mention in the Chronicle of Higher Education's new books section?
    - promote the book at a nation conference?
    - organize a series of book readings?

15. For promotion and tenure, it is essential to have publications based on research done at your current institution. Have you set goals for completing research and writing in time to publish the results before critical review dates? Have you talked with your mentor(s) about these expectations?

16. Performing reviews will help you to develop the skills to effectively evaluate your own work.[3] Consider containing the editor of key journals or publishers in your field to indicate your willingness to act as a reviewer. Include information about your area of specialization.

2.4.5   MENTORING ACTIVITY: THE MANUSCRIPT SUBMISSION AND REVIEW PROCESS

Most publishers use a similar process for the review of a manuscript for publication. The schematic[4] below provides a general sequence for how this often occurs. Work with your mentor to choose a publisher for the next manuscript you plan to submit and review the guide to authors (often available on their website). Is the process depicted below accurate for this publisher? How should it be modified?

[3] A.J. Smith, "The Task of the Referee," **Computer**, p. 65-71, April 1990.
[4] Adapted from: K. Barker, **At the Bench: A Laboratory Navigator,** Updated Edition, Cold Spring Harbor Laboratory Press, Cold Spring Harbor, NY, 2006.

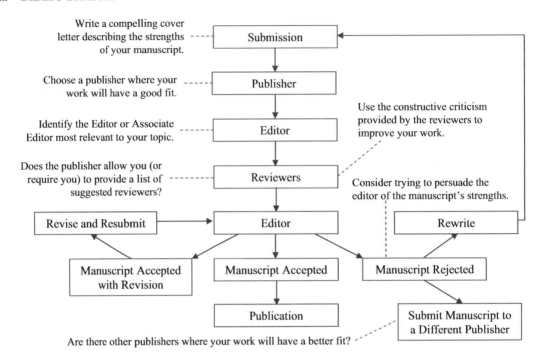

BIBLIOGRAPHY

JOINING YOUR DEPARTMENT AND DISCIPLINE

R. Boice, **The New Faculty Member: Supporting and Fostering Professional Development**, Jossey-Bass Publishers, San Francisco, CA, 1992.

E. Evans and C. Grant, editors, **Mama, PhD: Women Write About Motherhood and Academic Life**, Rutgers University Press, Piscataway, NJ, 2008.

"Facts & Figures," The Chronicle of Higher Education, `http://chronicle.com/section/Facts-Figures/58/` (Accessed 11/11/09).

W. Germano, **From Dissertation to Book**, University of Chicago Press, 2005.

G.W. Gibson, **Good Start: A Guidebook for New Faculty in Liberal Arts Colleges**, Anker Publishing, Bolton, MA, 1992.

E. Harman, S. McMenemy, I. Montagnes, C. Bucci, editors, **The Thesis and the Book: A Guide for First-Time Academic Authors**, Second Edition, University of Toronto Press, 2003.

R. Hile Bassett, **Parenting and Professing: Balancing Family Work with an Academic Career**, Vanderbilt University Press, Nashville, 2005.

B. Luey, **Handbook for Academic Authors**, Third Edition, Cambridge University Press, Cambridge, 1995.

B. Luey, editor, **Revising Your Dissertation: Advice from Leading Editors**, University of California Press, 2008.

C. H. Sides, **How to Write and Present Technical Information**, Oryx Press, Phoenix, AZ, 1999.

In a new academic position, there is much to learn in a short time. Looking back on just the first few years as an assistant professor, I find that I improved more in my teaching, learned more about my research area, and gained more skills through my service activities than I did in the entirety of my graduate schooling. I think many of us had the illusion that we would be prepared when we arrived on campus for our first academic job. However, in many ways, the experiences that come before your first academic job are just a taste of what comes next!

CHAPTER 3

Establishing Expertise

Teaching, research and service are the traditional triumvirate of academic life; however, in each institution, college, and department, the relative importance and weight of teaching, research and service will vary considerably. Thus, your academic career will be in some way divided between these three areas. In the best of all worlds, your teaching, research and service will also be integrated within themselves. It is your job to determine your own academic equation for success - both for your own personal satisfaction and for the academic setting you have chosen.

In some fields at some institution, you will be required to develop an international reputation in at least one of these areas. How you accomplish this and make an impact on your field will be unique however. Except in rare circumstances, you do not want to make yourself into a clone of your colleagues. You will want to develop your unique contribution to the field and the department. You will also need to strike a balance between being integrated with the mission of your department and developing your own distinctive contribution. The key is to make yourself indispensable.

At some institutions, it will be important to make substantial progress in all three areas prior to the first evaluation of your performance. You can do this successively (taking chunks of time for each area in turn) or simultaneously (the ultimate in multitasking). The relative amount of time and energy as well as the depth of expertise and external reputation you need to develop with your scholarly work will differ for each department and institution.

When considering the questions below about managing your professional life, ask yourself how you are addressing all three areas of this academic equation for success.

3.1   TEACHING

3.1.1   OVERARCHING QUESTIONS TO CONSIDER

- What resources are available to you for further developing you teaching skills?

- How have you evaluated your success in teaching?

- In what ways can you incorporate your research or service expertise into your teaching?

- How have you developed your advising and mentoring skills?

3.1.2   MENTORING CONVERSATION: ON TEACHING SKILLS

Recently, one of my colleagues told me a story about an interview with a faculty candidate. He had asked the candidate a question about how students learn. The candidate seemed stumped and replied

"I never really thought about how students learn." Of course, this was not the best reply to give for someone trying to obtain a faculty position.

The reality is, however, that most faculty have spent little time thinking about how students learn. Certainly, we think about our teaching, but, often, we do this thinking in a very self-centered way, focusing on how we were taught, how we came to understand a subject, and how we will teach *to* our students. This is not too surprising given the lack of attention that the subject of teaching traditionally receives in most graduate student education and postdoctoral training programs. Although a major percentage of a faculty member's job is focused on the topic of teaching students – with the intent that they actually learn something – traditionally, we are ill prepared for this function.

The good news is that graduate education is changing and postdoctoral training opportunities that focus on education (from many different disciplinary backgrounds) are becoming more available. I was lucky to have participated in a Preparing Future Faculty program as a graduate student so that I did not feel entirely adrift during my first full fledged teaching experience. However, training in the area of teaching and learning is not often a standard part of the graduate training experience. Even if it were, this is an professional area where lifelong learning is critical for faculty. Numerous workshops (both at your institution and offered nationally) and books on teaching and learning are available. You will likely find, as I did, that time spent broadening your horizons in this area is worthwhile, making your teaching, more creative, enjoyable, and effective.

Even if you can't find existing programs that suit your needs, you can make them happen. A few years ago some of the junior faculty in my college started a weekly morning coffee conversation that regularly addresses teaching issues, which is still going. Continue to seek out training workshops and conversations with colleagues about teaching, because, as in research, there is always more to learn, explore, and test!

3.1.3   DETAILED QUESTIONS AND SUGGESTIONS TO CONSIDER

1. Teaching a course for the first time is exceptionally time consuming. Consider:

   - Talking to colleagues about the syllabus, texts, and reading lists they have used in the past

   - Drawing upon the examples of your own teachers

   - Sharing ideas and information with untenured colleagues at peer institutions

2. The tone is set for a course in the first week or two in multiple ways and is difficult to shift well into the semester. One aspect of this is how the students address you. You will need to give them guidance on how to address you appropriately in person and in email. Do want to be called professor, doctor, or by your first name? What is the common practice in your department? How will your choice affect student interactions? You might consider different choices depending on whether the students are undergraduates, graduates, or members of your research group.

3. Developing a new course may be strongly encouraged to enhance your tenure case. If this is not required, is this something you are interested in doing? Is it wise to spend your time in this way? Particularly, if the course topic is very unique, it may be a very time consuming process. Can you facilitate new course development by:

   - looking for similar courses offered at other institutions and requesting information such as the syllabus and reading list form the instructor?

   - seeking new textbook(s) related to the topic?

   - approaching other experienced teachers for advice on new course development?

   - ensuring that your chosen topic does not encroach on the territory of other faculty in your department or institution?

   - speaking with your department chair or director of undergraduate/graduate studies about when your course might be incorporated into the timetable

   - acknowledging to yourself that it? will not be perfect the first time you teach it?

4. What resources are you using for teaching development? Consider participating in the following:

   - Faculty development seminars
   - Teaching portfolio workshops
   - Courses/workshops on teaching pedagogy and curriculum development in your field
   - Committee work on curriculum development

5. How can you best document your teaching effectiveness through your development as an instructor? Consider using some of the following evidence:

   - Standard student evaluation forms
   - Feedback from surveys of graduating seniors or alumni
   - Letters from students who have taken your class
   - Letters from colleagues who have observed your teaching
   - Letter from a teaching improvement specialist who has interviewed your students or analyzed the content of your instructional materials

6. Student evaluations of your teaching can be a helpful resource for improving your teaching, are often used as a measure of your success as a faculty member, and may be included in your tenure packet. There are a number of issues related to teaching evaluations that you may want to consider:

   - Numerical rankings are compared to departmental and college averages – do you have access to these average rankings?

- Often evaluators are looking for a positive trajectory with teaching improvement - are you tracking your teaching evaluations over time?

- Some written comments are inappropriate and should be ignored – can you let go of the hurtful comments?

- Some suggestions may run counter to your teaching philosophy – did you explain your philosophy to your students?

- Student expectations of male and female instructors are different – does your teaching style run contrary to gender norms? Could this explain some of the comments you get?

- Sometimes it is difficult not to take students' comments personally – students are used to being evaluated critically; faculty have to get used to it too!

7. Peer evaluation of teaching by an experienced faculty member is critical tool for improving your teaching and also is used as a measure of you success as a faculty member. In some institutions, peer evaluation of teaching is a requirement for your tenure case. How you will facilitate making it happen? How you will make it as productive and constructive as possible? How it will be documented for your tenure file? Have you considered:

- requesting that someone observe your teaching through your department chair, mentor committee, and/or teaching center at your institution?

- contacting the person who will observe you to arrange a date?

- arranging the observation for a class in which you feel confident about the material or want to get feedback when you plan to try something new?

- providing the observer with a description of the course, syllabus, overview of what your plan to do in that class and how it fits into the context of the overall course, readings or assignments to be covered in the observed class, and any handouts for the class period observed?

- being well prepared but not over prepared?

8. What expertise are you developing within your discipline and in related disciplines? How do the classes you teach help you in developing that expertise? Can you think of ways to improve this? For example, you might:

- talk to faculty who teach in your areas of interest.

- talk to other faculty about how best to juggle teaching multiple courses in the same term.

- teach seminars about your area of interest.

- present at conferences and association meetings.

9. In what ways can you incorporate your research expertise into your teaching? Consider:

- Using examples from your research to provide a larger context
- Choosing visually appealing or tactile aspects of your research for introduction in an undergraduate classroom
- Incorporating small research oriented projects as a course requirement
- Using a class project as a way to refine or test an aspect of your own research projects
- Developing a new course related to your research in order to provide yourself with a learning opportunity

10. A person's teaching style is often informed by their experiences as a student.

- In what ways does your own experience as a student influence your teaching?
- How are students in your class different from who you were as a student?
- Are the teaching methods you use effective with the diverse set of students in your classroom?
- What can you do to teach more effectively to a broader range of students?

11. Does the way you think about teaching and how you construct your courses involve the whole student?[1] Have you considered how you can:

- foster self-authorship[2] in your students and promote lifelong learning?
- provide genuine links between students' diverse lives and experiences and academic learning?
- prepare students to meet the demands of a professional and adult in society?
- prepare students to actively engage in bettering the world around them?

12. Teaching involves a variety of different settings and communication styles. Consider how you would accomplish the following:

- Facilitate student participation in the classroom
- Incorporate group projects
- Handle a large lecture class
- Help your students to engage in active learning
- Obtain feedback from minute papers and informal evaluations
- Evaluate student understanding effectively

---

[1] M.B. Baxter Magnolia, **Creating Contexts for Learning & Self-Authorship: Constructive- Developmental Pedagogy**, Vanderbilt University Press, 1999.

[2] From Baxter Magnolia, p. 10: "Self-authorship is simultaneously a cognitive (how one makes meaning of knowledge), interpersonal (how one views oneself in relationship to others), and intrapersonal (how on perceives one's sense of identity) matter."

- Coach your teaching assistants

13. There are a number of issues that are not covered in the textbook for your course but still affect the learning process. You need to prepare yourself to handle students who come to you with issues concerning the following:

- Dropping your class
- Exam anxiety
- Late work
- Illness or death in the family
- Grade arguments
- Complaints about your teaching style
- Sexual harassment

14. Documenting your teaching is also important. How well developed is your teaching portfolio? How have you evaluated your success in teaching? Your teaching portfolio[3] should include the following:

- Statement about your teaching philosophy
- Reflections on your teaching experiences
- Student evaluations
- Peer evaluations
- Syllabi you have developed
- Seminars/workshops attended
- Training programs and certifications
- Records of your committee work related to teaching
- Plans for your future teaching

3.1.4   MENTORING ACTIVITY: DEVELOPING STRATEGIES TO CREATE INCLUSIVE CLASSROOM

Step 1: Consider the teaching methods and instructional choices identified below and identify who might feel excluded when they are used.[4]

Step 2: Identify at least one strategy for each area which will make the classroom more inclusive.

---

[3]"Teaching Portfolios," Teaching and Learning Excellence, Teaching Academy, University of Wisconsin – Madison, https://tle.wisc.edu/teaching-academy/peer/tfolio (Accessed 11/11/09).

[4]Adapted from: J. Handelsman, S. Miller, C. Pfund, **Scientific Teaching**, The Wisconsin Program for Scientific Teaching, W.H. Freeman and Company, New York, NY, 2007.

Step 3: Identify other teaching methods you regularly use in your classroom and complete steps 1 and 2 for these.

Step 4: After drafting your strategies, seek out input from peers and mentors about other strategies to improve inclusion in the classroom.

Teaching Method or Instructional Choice	Who might feel excluded?	What could be done to make the classroom more inclusive?
Lectures are done with PowerPoint exclusively. The slides are dense with information and are not available to students.		
Historical examples always involve white men.		
Examples always involve people and situations from the U.S.		
Exams are entirely multiple choice and true/false.		
Exams are timed. Students are cut off after 1 hour. Grades are based on a curve.		
Homework assignments are only available online.		
The textbook costs $150.		
The class meets at 7:30 AM.		
The class meets at 7:30 PM.		
The course requires students to work in assigned groups outside of class time.		

3.2   ADVISING AND MENTORING STUDENTS

3.2.1   OVERARCHING QUESTIONS TO CONSIDER

- Can you broaden your view to take into account the breadth of backgrounds and aspirations that your advisees will bring to the table?

- Have you established clear expectations for the roles in the relationship with your advisees/mentees?

3.2.2   MENTORING CONVERSATION: BEYOND ADVISING

There will come a time when one (if not many) of your students will penetrate the barrier of professional distance and become important to you in a way that is closer to how you might feel towards your own child or a family member. You become concerned about their well being and are invested in their success. This is not necessarily a bad thing, but it is something that you may want to prepare yourself for. You may also want to identify boundaries for yourself - certainly, you would not want to do anything that could even be perceived as sexual harassment - but, that aside, you also may want to limit how close you allow yourself to become to students who you advise and mentor. As with any relationship, there are opportunities for disappointment and hurt feelings.

Yet, the risk is very worthwhile. I have had the opportunity to help the flunk-out-freshman turn his life around when he become a father, support the stellar student after the death of a parent, encourage the all-around-achiever who was enthusiastic about topics I am passionate about, and mentor several through the process of obtaining a faculty position. There are a number of students who I have developed strong mentoring relationships with that have even developed into friendships over the years. We keep in touch by email or phone, and some, I see at conferences. An important thing to remember is that the students "grow up," and your relationship will need to change over time.

Although I have found advising and mentoring students to be incredibly rewarding, I have also been disappointed at times – both in my ability to effectively guide students and their ability to live up to their potential. The part I struggle with the most is advising students who are not on a path they are cut out for or interested in. It's a tough situation because there is sure to be disappointment - either for them or the family members who are pushing them to do something they are not interesting in doing. But certainly, the disappointment is greater when someone has wasted even more time traveling down the wrong path farther than they should. Handling each situation is different, but I would advocate that it must include a conversation where you are honest and forthright about what you see as their strengths and weaknesses and how those do not fit with the career path that they have embarked on with their studies.

Harry Levinson summarizes prior theories of management which propose that "People are most deeply motivated by work that stretches and excites them while also advancing organizational goals."[5] He goes on to point out to managers that they must evaluate the individual and the orga-

---

[5] H. Levinson, "Management by Whose Objectives?" **Motivating People (Harvard Business Review**, January 2003.

nization and determine if there is a significant discrepancy. If so, "The person might well be better off somewhere else, and the organization would do better to have someone else in place whose needs mesh better with the organization's requirements." Thus, you may need to guide your student through introspection and self-evaluation to determine the magnitude of the discrepancy. In some cases, however, an individual may not be able to do this, and you may have to identify the discrepancy and take action to modify or terminate the relationship. But, I would suggest that you look into procedures and restrictions with your department administration before embarking on termination.

I have been disappointed with students who are not living up to their potential, and I have run into situations where I had to decide whether it was better to push and be critical or back off and let them find their own way. However, even when a mentoring relationship appears to have failed, there may have been a positive influence made on the student that we did not realize. If we are lucky, they will reconnect later on and tell us.

3.2.3 DETAILED QUESTIONS AND SUGGESTIONS TO CONSIDER

1. How do you advise and mentor students? Consider the differing needs and expectations of people with whom you interact:

   - An undergraduate student in your department
   - An undergraduate research assistant
   - A graduate teaching assistant
   - A graduate research assistant
   - A postdoctoral researcher

2. In some departments, faculty are assigned a group of undergraduate students to directly advise, while other departments use group advising or a staff member advisor. Even if you are not assigned a group of advisees, you may have undergraduate students you work with more closely and you may have graduate students that you advise. As an advisor, there are aspects of your institution that you will want to become familiar with in order to most effectively guide with your advisees:

   - Degree requirements
   - Major/minor options
   - Class schedules
   - Registration and drop dates
   - Study abroad opportunities
   - Internship/coop opportunities
   - Career services
   - Writing services

- Tutoring services

- Learning disability services

- Counseling services

- Harassment policies

- Threat assessment and early intervention

3. Although it may seem constraining, there are a number of suggestions to consider in order to avoid any allegations of sexual harassment when meeting with students and to avoid being the victim of sexual harassment:

- Set clear boundaries about behavior.

- Always meet with an open door.

- Avoid physical contact with a student.

- Avoid language that could be construed as lewd or suggestive even if it is intended to be humorous.

- Never suggest that a grade or letter of recommendation is dependent on anything other than academic performance.

- Do not display suggestive or pornographic pictures.

- End a meeting and/or leave your office if you feel uncomfortable.

- Discuss potential problems with your chair or senior colleague before they grow.

4. In the wake of campus violence, many institutions have developed threat assessment and early intervention policies. There are some things that you can do in your one-on-one interactions with students to manage your interactions with troubled people:

- Set clear boundaries about behavior.

- Remain calm and do not lose your composure.

- Don't ignore warning signs.

- End a meeting and/or leave your office if you feel uncomfortable.

- Contact the institution's threat assessment team, counseling services, employee assistance office, or dial 911.

5. What do your advisees expect from you? Are you giving them:

   - constructive feedback?
   - assistance in setting realistic goals?
   - feedback about expectations?
   - information about funding opportunities?
   - professional development opportunities and connections?
   - assistance in preparing a resume or CV?
   - aid with the job search?

6. Can you foster creativity[6] by providing:

   - challenge?
   - freedom?
   - resources?
   - a supportive workgroup with a diversity of perspectives and backgrounds?
   - collaboration and information sharing?
   - encouragement?
   - protection from political issues?

7. Have you considered developing a contract with you advisees outlining expectations on both sides?

   - Propose a regular meeting schedule.
   - Define your expectations for communication/reporting.
   - Identify which issues can be decided on independently and which must be done in consultation.

8. Developing trust with your students, advisees, and mentees is an important goal. Robert Hurly[7] suggests that the decision to trust depends on that individual's risk tolerance and your ability to accomplish the following:

   - Spend time explaining options
   - Explain decisions
   - Recognize achievement
   - Coach through failures

[6] T.M. Amabile, "How to Kill Creativity," **Harvard Business Review**, September-October 1998.
[7] R.F. Hurley, "The Decision to Trust," **Harvard Business Review**, September 2006.

- Demonstrate your skills
- Delegate tasks appropriately
- Display predictable behavior
- Act with integrity
- Communicate with candor
- Emphasize your commonalities
- Demonstrate concern
- Cultivate bonds

9. In what ways have you informed yourself about student populations, opportunities, and problems at your institution?

10. Do you have experiences teaching students from less academically and economically advantaged backgrounds or students with disabilities? In what ways are you developing sensitivity to issues of multiculturalism?

11. Most students will not follow the path that we took. This is particularly true for the undergraduate advisees who are assigned to you and often have a wide range of long term goals.

  - Have you asked your advisees about their plans, goals, and dreams, as well as their struggles, disappointments, and failures?
  - Have you offered them information about a range of possibilities given their interests?

12. How do you envision your role as mentor to your graduate students, trainees, and post docs? How will you help your students to:

  - develop into a junior colleague?
  - work towards professional independence?
  - develop an understanding of how their work fits into the broader context of the field?
  - connect with researchers in their field of study?
  - develop their communication skills?
  - be inspired to become mentors themselves?

13. In some cases, your role may be that of a supervisor rather than an advisor or mentor. Consider how you can structure the work functions[8] of someone reporting to you by providing them with the appropriate amount of:

  - variety,

[8] J.R. Hackman and G.R. Oldham, "Motivation through the Design of Work: Test of a Theory," **Organizational Behavior and Human Performance**, 16, p. 250-279, 1976.

- flexibility,

- meaningfulness,

- autonomy with responsibility, and

- feedback with appreciation.

14. You will likely be asked to write letters of recommendation for your students. These letters usually comment on:

  - how you know the student and for how long,

  - the student's work ethic and productivity,

  - examples of their creativity and independent thinking,

  - the student's ambitions for the future, and

  - their ability to work effectively with others.

  A strong letter of recommendation should include at least one detailed example of the student's performance or strengths.

15. There are a number of challenging questions that may arise as you mentor students. Have you considered:

  - What advising strategies will you use when your student's goals differ from your own?

  - How involved should you become in your student's research project?

  - How involved should you become in your student's personal life?

  - How close can the relationship become before you lose your ability to objectively evaluate your student?

  - How might differences in gender, sexual orientation, race, ethnicity and nationality affect your mentoring relationship?

3.3   RESEARCH AND SCHOLARLY ACTIVITIES

3.3.1   OVERARCHING QUESTIONS TO CONSIDER

- What are your research goals and what methods will you use in your approach to achieving these goals?

- How can you acquire the resources you will need to be successful in your scholarly work?

- What skills will you need to be successful in obtaining research grants and faculty fellowships that exist in your field?

3.3.2   MENTORING CONVERSATION: ON THE UPS AND DOWNS OF DOING CREATIVE WORK

Your pursuit of scholarly work, i.e., your research, can be both invigorating and demoralizing. Sometimes your creativity abounds and new exciting ideas jump out at every turn; your ideas receive positive review and praise in your research community; and your interactions with colleagues, collaborators, and students fuel your enthusiasm to such an extent that you feel you can solve all the problems of the world. Unfortunately, there are other times – I'll call them slumps – when the well appears to be dry and it is unclear to you if you will ever have an original creative thought again.

When we are honest with each other, most everyone will admit to having *both* the ups and the downs. The problem is that you seldom hear people talking about the slumps. In fact, most academic cultures promote secrecy and active denial of the occurrence of slumps. Possibly, this strategy is based on the hope that if we don't talk about it the slumps won't occur, or if we ignore them, they will go away. Or it could be that the academic culture has a survival of the fittest mentality, so to show weakness would put you in danger of being culled from the pack.

I have come to believe that these ups and downs are a natural part of the cycle of any career that demands creativity. Smart creative people can't be smart and creative 100% of the time. I also believe that the denial we actively engage in often exacerbates the problem. So, I'll admit it to you, I have had slumps. For the most part, I too have hidden them from my colleagues. For me, I think it is primarily the fear of being accused of being a fraud – the old "imposter syndrome." But what I have discovered over the years is that, if I can find one trusted colleague that I can feel comfortable talking with honestly and assured in their ability to keep my confessions confidential, the conversation relieves much of the burden and is often enough to turn the tide and give me the means to get myself out of the slump. When I have initiated these conversations, I have also found my colleague telling me "I've been feeling the exact same way recently" or "I remember feeling the same way you are now back when…." Knowing you are not alone also relieves some of the self doubt.

We all might be a bit more balanced if we could be more open and honest with our colleagues about our successes as well as our failures. Often, the failures are the more enlightening learning experience and the slumps are just temporary. But I would still caution an untenured faculty member to hold back things that could be perceived as negative from those who have decision-making authority over their careers. Unfortunately, not everyone is understanding, and some will even use your weaknesses against you for their own political or prestigious gains. A former colleague of mine found this out the hard way after admitting an area of weakness to a senior colleague while seeking help to improve. This colleague ended up using that weakness as the primary evidence as to why the department should not reappoint her. Although very happy and successful in her new career, she is no longer on the tenure track at that academic institution.

3.3.3   DETAILED QUESTIONS AND SUGGESTIONS TO CONSIDER

1. Different institutions require different levels of engagement in research, from none to substantial.

- Will you be expected to do research with undergraduate and/or graduate students?

- Are postdoctoral students common?

- Will you be able to focus on research throughout the year or only at term breaks?

- Will your research be conducted on campus, in the field, or at another institution or facility?

- Have you determined whether or not developing an "international reputation" is expected for tenure at your institution?

2. What resources are available for research development at your university? Consider how you can benefit from participating in:

  - programs specifically designed for new faculty,

  - research seminars,

  - grant writing or grant management workshops,

  - committee work on research issues, and

  - preparation of a research portfolio.

3. If your research involves a laboratory setting, how quickly can you get it set up so that you can start producing results? Can you hire someone to help you order and set up equipment, consider an undergraduate student or postdoctoral researcher? Consider whether you can use equipment from a shared facility or someone else's laboratory to get results while you are working on yours.

4. Additional skills are required to manage a laboratory and/or research group. Identify ways in which you are developing expertise in the following:

  - Recruiting and managing personnel

  - Managing grants

  - Budgeting research projects

  - Purchasing

  - Understanding university policy and procedures

  - Overseeing health and safety policies

  - Managing animal and human subjects

  - Understanding regulations pertaining to research human subjects

5. How will you develop research expertise if you are in an ineffective research setting? How can you maximize opportunities that are available to you? Are there other universities or facilities in your area where you can conduct research? Consider:

- Research at a science or history museum
- Research with archival and special collections
- Consultation with partners in business and industry
- Technology transfer programs between your university and industry
- Collaboration with national laboratories
- Short-term research symposia at other universities and organizations
- Short courses
- Research abroad opportunities
- Time away from teaching

6. In addition to teaching, research and service abilities, you need a whole set of ancillary skills. What resources are available, and who can help you develop these skills? Consider how you will gain experience in the areas of:

- writing and editing,
- communicating effectively with others,
- networking,
- negotiating, and
- time management and goal setting.

7. Have you developed a research portfolio? Such a portfolio would include the following:

- Descriptions of your research projects and explanation your research's impact on your field
- Reprints of journal articles authored
- Committee experiences

8. Evaluate your research goals. What depth and breadth of research methodology and scholarly experience will you need? How are you developing depth and breadth in your field? What can you do to expand your experience?

9. How do your current research projects expand or limit your opportunities for the future? Can you think of ways to create future opportunities? Consider:

- Talking to senior faculty about developing your research program
- Talking to other faculty about how they developed expertise
- Talking to senior faculty about how they would choose an area of expertise today
- Teaching seminars about your areas of interest

- Presenting your research at conferences and association meetings
- Networking/collaborating with other investigators, including your peers and researchers you meet at meetings and conferences

10. It is often important to separate yourself from the research you conducted with your former advisor(s). How can you distinctly frame your research within your field? Can you use the skills you developed in your prior research experience while venturing into a new area of research?

11. How broad should you be in your scholarly pursuits? What is an appropriate number of research projects to develop that will allow you to be diversified while still maintaining progress?

12. There are a number of different approaches you can take in building your research program. These are dependent on your own desires, your discipline, and your institutional setting. Have you considered the pros and cons of the following strategies?

    - *The Lone Wolf* - develops an independent research program, applies for single investigator grants, works with a small number of students and post doctoral fellows.
    - *The Empire Builder* - develops and manages a large research group, applies for large single investigator grants, employs a large team of students, post doctoral fellows, and research staff and/or advises several dissertation students and teaching assistants.
    - *The Sequential Collaborator* - identifies and builds collaborations with different colleagues, applies for both single investigator and multiple investigator grants, works with a small team of students and post doctoral fellows that may be co-advised.
    - *The Team Player* - develops or joins a collaborative research team with multiple faculty investigators, applies for multiple investigator and center grants, works with a large team of students, post doctoral fellows, and research staff .

13. Developing a successful research group is dependent on the people you have working for you. How do you recruit talented students to your research group? Consider:

    - Keeping an up-to-date web page
    - Participating in departmental recruiting activities
    - Inviting the best undergraduate or graduate student(s) in your class(es) to join your group
    - Mentoring undergraduates in summer research programs
    - Involving yourself in the graduate student admissions process
    - Meeting with new graduate students in the fall
    - Teaching courses taken by first year graduate students

14. At some institutions, it is possible to take an untenured leave or arrange your duties with your department such that you have a semester devoted to research. When is an opportunity such as this best timed? How will you most effectively use your time to your benefit? Work with your mentors to determine these answers and discuss whether your time should be more or less focused on conducting scholarly work, writing and publishing research products, or writing and submitting proposals to fund your work in the future.

3.4 GRANTS AND FUNDING

3.4.1 OVERARCHING QUESTIONS TO CONSIDER

- What level of support will you need to establish and maintain the type of research program you envision?

- Have you sought out advice and experiences that will aid you in the process of identifying where grant funding is available an how to go about obtaining it?

3.4.2 MENTORING CONVERSATION: ON WRITING YOUR FIRST PROPOSAL

In a recent summer, I gave my fifth annual lecture to junior faculty about "Examples of Successful CAREER Awards." This is part of a two-hour workshop that the university provides. It is a great workshop that helps train people about this particular award and the funding opportunities available from the National Science Foundation in general. It is also one of the key reasons why our faculty have been so successful in obtaining CAREER Awards in recent years. But I also feel sorry for these would-be award winners who are embarking on the proposal writing process. For many of them, it is the first proposal they will write in their academic careers. There is a lot of pressure to do it well and even more anxiety about "what if I fail." The workshop itself is a bit overwhelming because the participants are bombarded with things they should do and think about for nearly two hours straight from a slate of faculty members who themselves have been successful at obtaining CAREER Awards or, in the case of the most senior speaker, a Presidential Young Investigator Award.

Even if your research area does not make you eligible for this particular award, there is likely an equivalent sort of young investigator award in your field which will supposedly "guarantee" that you get tenure if you can win the award. It is undoubtedly highly competitive with a seemingly ever higher set of hurdles that you must jump in order to win it. The best advice I can give you is: apply, but do not expect to win. Hopefully, this approach will take the edge off of the anxiety.

If you don't win, you still have not wasted your time. You learned about proposal writing and aspects of your university infrastructure designed to help (or in some cases hinder) you in the process. The next proposal you write will be easier because of the experience. The intellectual effort is also not wasted, these ideas can be refined and used again in another proposal. My first attempt to get a CAREER Award was unsuccessful. But I got useful comments from the reviewers, refined and expanded the proposal, and submitted it to a different funding agency. This proposal did get funded

– and because of the inclusion of several collaborators – at a funding level four times greater than the original proposal. The next year, I took what I had learned about proposal writing and the nuances of the selection criteria peculiar to the CAREER Award and developed a new proposal based on new research and education ideas in another related area. The practice I gained the previous year paid off, and I was granted the award.

There is a lot of advice I and your colleagues can give you about writing proposals. Some of the questions below can prompt you to seek out more detailed guidance relevant to your discipline and the particular funding you are applying for. The most important advice I can give that applies to everyone in every field is that you should only write proposals on topics you are genuinely excited about pursuing. Without this basic ingredient, the proposal writing process will be painful, you are not likely to get the funding, and if you do get funded, you will be stuck working on something you are not really interested in. Lack of enthusiasm is difficult to disguise, and if it shows through in your proposal, you will have no chance of getting the funds. Don't follow the hot trendy topics just because there is money there. Lots of people are chasing these topics already anyway, so the dollars to person ratio is likely to be low. Instead, be creative and find new ways to approach the topics you are really excited about and sell these ideas in the best way possible.

3.4.3 DETAILED QUESTIONS AND SUGGESTIONS TO CONSIDER

1. There are a number of things that you can get grants and funding for which will help you advance your career. Some of these items are more obvious than others:

   - Research equipment
   - Research project assistants
   - Summer salary
   - Archival research
   - Workshops and conferences
   - Travel
   - Course load reduction
   - Sabbatical leave
   - Teaching improvement projects
   - Teaching equipment and laboratories

2. What research grants and new faculty fellowships exist in your field? Do you review these opportunities regularly? What are the major funding agencies and foundations in your field? Check or consult with a range of sources:

   - The faculty in your department
   - Deans with responsibility for research in your college

- Deans of the administrative units such as the Graduate School and the Office of Research and Sponsored Programs

- Offices responsible for research technology transfer

- Government agencies

- Private foundations

- Grant agency web sites and computerized search programs such as GRANTS.GOV, SPIN, IRIS, and COS[9]

- Electronic notification services

- National associations and other professional groups

3. What resources are available on your campus for seed funding, travel grants, untenured leave, and graduate student funding?

4. What resources may help you apply for fellowships and grants? How can you find these resources? Consider utilizing the following resources:

- Books on writing grant proposals

- Examples of successful/unsuccessful grants in your field

- Grant writing classes and workshops

- Service as a reviewer for a granting agency

- Individuals with expertise in your research area who will critique your proposals

- Application guidelines and procedures

- Individuals at your institution who can offer assistance with budget preparation electronic grants submission, facilities descriptions, etc.

5. Grant proposals must be persuasive. When writing a grant think about three questions:

- Have I demonstrated that my ideas are creative and will make an impact in my field? (Sell the idea.)

- Have I demonstrated that I am capable of conducting research that I propose? (Sell your self.)

- Have I demonstrated that I am motivated to conduct the research that I propose? (Sell your enthusiasm.)

6. A grant proposal must also contain all of the necessary ingredients to be considered favorably. Consult the guidelines of the agency you are submitting to and ensure that you have included a:

---

[9]See bibliography for links to these tools.

- compelling introduction that addresses the impact of and the need for the research you propose,

- description of the research that shows that you have both a broad and detailed understanding of the field,

- polished literature review that includes citation of your own work (if applicable) and of influential investigators conducting research on the topic,

- research question that is of a reasonable size,

- discussion of the proposed research that identifies its uniqueness or novelty,

- clear outline of the research goals,

- preliminary results (if available),

- conclusion that sells the proposal - the last few paragraphs are as important as the abstract,

- list of your prior support from the agency, which includes publications and students resulting from each grant,

- description of broader impacts or educational component (if required), which supports such things as incorporation of the research into your teaching, training of graduate students, involvement of undergraduate research assistants, regular research meetings, or a track record of success with previous students,

- justification of all budget items,

- thorough explanation of equipment and travel needs,

- description of existing equipment and facilities available in your lab or at your institution, which will aid this research, and

- support letter(s) from your collaborator(s).

7. Reviewers will receive your proposal more favorably if it is readable and visually appealing. Consider the following suggestions:

   - Include images, schematics, and diagrams that will elucidate your ideas.

   - Reiterate and summarize your key points and goals throughout the proposal.

   - Use italics and bold fonts for key words, phrases, and sentences to make outstanding things stand out.

   - Make lists and numbered paragraphs to clarify your ideas.

8. One of the best ways to learn how to write better grants is to read examples of good and bad grant proposals written by other researchers in your field. Have you considered contacting a granting agency program manager in your research area to express your interest in serving on a proposal review committee/panel?

9. Part of a granting agency or foundation program manager's job is to provide you with information about the agency and the types of research that the agency is interested in funding. You can take advantage of this by:

   - calling the program managers in relevant agencies and foundations to discuss the types of research currently being funded by their program.

   - setting up a meeting to visit the program manager personally to discuss your research ideas.

   - sending the program manager a white paper describing your research idea so that you can obtain informal feedback.

10. Some departments/institutions give more weight to single investigator proposals in the tenure review process. Have you discussed the emphasis on single versus multiple investigator grants with your mentors/department chair?

3.4.4   MENTORING ACTIVITY: THE PROPOSAL SUBMISSION AND REVIEW PROCESS

The specifics of what must be included in a proposal for funding differ depending on the agency and the funding for which you are applying; however, the general submission process is usually the same. The schematic[10] below provides a general sequence for how this often occurs. Work with your mentor to choose an appropriate funding agency and funding program for your planned research and review the proposal submission guidelines (often available on their website). Is the process depicted bellow accurate for this agency? How should it be modified?

3.5   SERVICE

3.5.1   OVERARCHING QUESTIONS TO CONSIDER

- How can you find out about the service expectations of your department, college, or institution?

- Can you identify service activities that will benefit you?

3.5.2   MENTORING CONVERSATION: ON ENGAGING IN THE RIGHT KIND OF SERVICE ACTIVITY FOR YOU

Multitasking is a fact of life for today's faculty members. The days of sitting alone in your office working on your scholarly pursuits for days on end or taking a six month field excursion are nearly nonexistent. A faculty member's job – no matter the type of institution – involves fulfilling a variety of obligations and wearing multiple hats. Particularly, in the days of lean budgets, faculty are relied

---

[10] Adapted from: K. Barker, **At the Bench: A Laboratory Navigator,** Updated Edition, Cold Spring Harbor Laboratory Press, Cold Spring Harbor, NY, 2006.

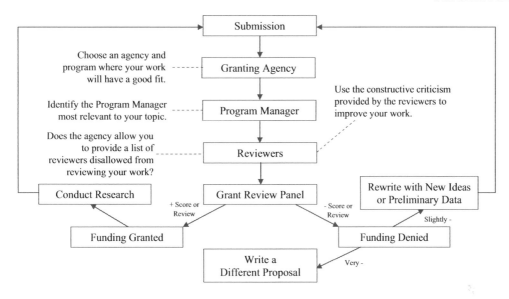

on even more to perform functions that were previously handled by staff. Staff are also overburdened and often performing functions previously handled by multiple people. I am personally very unhappy with this trend, but today, it is a reality at many institutions.

Something that can benefit you when it comes to keeping all of the various expectations under control is efficiency. Particularly, for those necessary but annoying tasks, you need to seek out strategies that help you get things done with the least amount of time and lowest aggravation level possible. This may mean compartmentalizing these tasks to certain times of the day or week. Or it may mean developing short cuts or templates for recurring tasks. It is also important to realize that spending twice as much time to make something just a little bit better is probably not an effective use of your time. You will have to determine when the level of completeness/perfection is good enough – this advice holds for research and teaching, and service as well day to day administrative tasks and flowing of email.

Although many faculty members would like to focus only on their research and teaching, service is a component of nearly every faculty position. Some obligations that fall into the category of service may feel like annoying tasks to you, but you may also be able to use them as learning opportunities. For instance, you might take on a role in your professional society organizing a symposium or workshop to increase your visibility in your field. If you do not already have good organizational skills, these opportunities will provide you with professional development in that area. It's important to also seek out advice from your peers and mentors to get helpful tips, the inside scoop on how things get done within the organization, and ideas for the best people to invite (those who will show up and do a great job in their invited role). Fundraising is also often a part of this process and these opportunities will help you to identify where there may be small pots of money for conferences or meetings that you can gain access to and grant opportunities that can help you to develop a relationship with program managers and grant officers.

There are many obligations that fall under the category of service that will be a benefit to you and, if you are strategic about it, they may even be fun! In many departments, there is some leeway in what service duties you are assigned. Choose ones that you can use to your benefit. For instance, serving on the graduate student recruiting committee would give you the opportunity to get a first look at all of the prospective graduate students, build relationships with them early on, and maybe even woo them to your group.

You may choose to take on other service duties because of personal interest. Anyone looking at my tenure packet would have told me – and did tell me – that it was a bit heavy on service. I took on a large service obligation after I was on campus for just a few years because I was really excited about the opportunity and I knew that I would find spending time on it very energizing. I will admit that my mentors advised against me doing this. I listened to their advice, but I decided to do it anyway. I felt that this sort of service role was one of the key attractions for being in an academic position rather than following a different career path. I also felt that it was something I wanted to take on and try to balance with my research and teaching obligations. If I could not pull it off and were to be denied tenure, then it was probably a good sign that I should not be pursuing this sort of career path. As it turns out, I got tenure and that service role I performed ended up providing me with a lot of unanticipated benefits.

3.5.3   DETAILED QUESTIONS AND SUGGESTIONS TO CONSIDER

1. What committees and other university service opportunities are available to you? Have you considered all three levels of the academic hierarchy – departmental, collegiate, and university? Based on your overall goals, have you assessed which committees will be most beneficial to your career? Possibilities include the following:

   - Service on search committees
   - Policy and review councils
   - Faculty governance committees
   - Advising undergraduate programs or organizations

2. What can you realistically expect to give in terms of time and effort toward service activities? Can you attend most scheduled meetings? What do you hope to gain from this experience? What committees have power? What committees would help you to broaden your network in the institution?

3. Women and minority faculty members are often called on to do a disproportionate amount of committee service. Do you assess each opportunity for its impact on issues you are concerned with, and its impact on your career? Can you resist taking more than your fair share of committee assignments?

4. Do you know how to be an effective member of a committee? What skills are you developing towards this goal? Consider trying to gain experience in the following areas:

- Lead a meeting

- Create/follow an agenda

- Participate in team building

- Facilitate a discussion

- Be heard in a meeting when you are outranked and overpowered

- Understand committee/meeting etiquette

- Prepare in advance to use time efficiently

- Make your comments succinct and meaningful

- Deal with conflict effectively

5. When and how will you gain experience in administration and leadership? In what ways are you challenging yourself to become a better group member, leader, or public speaker?

3.6   THE BALANCING ACT

3.6.1   OVERARCHING QUESTIONS TO CONSIDER

- In what ways will you balance teaching, research, and service roles?

- Do you know how and when to say "No"?

- How are you balancing your professional responsibilities with your personal life?

3.6.2   MENTORING CONVERSATION: ON USING ONE OF THE FIRST WORDS YOU LEARNED AS A TODDLER

Saying "No" and knowing when to say "No" are skills that often must be learned. Some people walk in the door with these skills or with other attributes that keep them from being put in situations where they should say "No." Many of us, however, are not so skilled in this arena. Much of the problem can be probably be traced to the sorts of personalities that end up being attracted to academic careers. The 'A' student is often seeking external approval and validation. These students want to please the teacher through excellent performance and the ability to go above and beyond the rest of the class.

But now as a faculty member and a new assistant professor there are many people to please. Many of your senior colleagues have a voice when it comes to tenure time, the dean of your college is likely to be part of the approval process too, and surely disappointing any one of them seems to spell doom, right? Wrong - this way of thinking is a trap. The reality is that if you agree to do everything that is asked of you, it is likely that you will be an ineffectual faculty member. You won't have time to get the things done that are critical to your position.

In most instances saying "No" to one specific request will not be held against you in perpetuity. It probably won't be held against you for even the blink of an eye. The requester will move on and try to figure out who to ask next, if they have not already constructed their list of "victims," which you have now been successfully scratched off of.

The key is that you must say "No" at least sometimes. So you need to have a set of criteria for accepting and rejecting requests and a strategy for how you will tell someone "No" thought out in advance. For a while, when one of my colleagues or the chair of my department asked me to do something, the first word out of my mouth was "No." It is really a rather blunt strategy and probably not the optimal one that I would advise others to adopt. The problem is that I had been struggling with being completely overloaded with both professional and personal responsibilities. So at that point in my career and life, the immediate "No" response was probably the wise one for me. I will admit that often "No" is not my final answer. The department chair quickly caught on to that fact and knew that in some cases I could be talked into it. The key was that I already have a set of criteria, so if he was able to meet those criteria then sometimes the "No" could be turned into a "Yes."

In previous years (before receiving tenure and later embarking on this blunt phase), I did not say "No" immediately, but I had also learned not to say "Yes" immediately. My strategy was to start asking questions to determine the nature and scope of the request, so I could figure out if it is something I should or want to say "Yes" to. For instance, travel was becoming a big problem for me a few years ago. I found myself away on travel far too frequently. It was disrupting my teaching, interactions with my graduate students, and my personal life. So I made a decision to cut back on travel. Some of this could be decided in advance – choosing which professional meetings were the most critical for me to attend – but other travel seemed to "pop up" through requests from external sources. My strategy was to make a list of questions to ask myself when considering new travel opportunities. The list that follows poses questions that I consider when faced with an invitation or a new opportunity that will involve added travel.

- What is in it for me?

- Is the purpose core to my goals?

- Could someone else do it instead?

- Could I postpone it until later?

- Will I have to miss teaching or other important things already on my schedule?

- Do I have other travel scheduled within the two weeks before or after?

- Can I put off making the decision until I know my calendar better?

- Is the location somewhere that I can tack on extra days to visit family or friends?

These questions may not be the right set of questions for you, but my advice to you is to develop your own set of criteria, stick to them, and find a way that you will feel comfortable saying

"No." You may have to have several sets of criteria for handling things like travel, committee service duties, student mentoring and advising requests, outreach and public interaction requests, and …. the list goes on.

3.6.3   DETAILED QUESTIONS AND SUGGESTIONS TO CONSIDER

1. In what ways will you balance teaching, research, and service roles? How would you describe the importance that your department and institution place on these roles? Where do your priorities lie? If there is a mismatch, have you considered how you will reconcile the differences?

2. Do you know how and when to say "No"? What strategies have you developed to avoid saying "Yes" when you want to say "No"? Specifically, can you turn down additional work yet maintain a positive relationship? Can you enlist the assistance of a senior colleague or mentor to help you determine when to say no and/or intervene on your behalf?

3. Performing scholarly work, like setting up laboratory experiments, analyzing data, writing manuscripts, and writing grant proposals, requires a depth of concentration that is often difficult to achieve when being constantly interrupted by administrative tasks, meetings, and student questions. Do you set aside blocks of time for research and writing? How do you guard this time against encroachment by other activities and ensure that you will not be interrupted? Do you make notes for yourself on where to pick up and what to do next when you stop working on something?

4. Frequently, part of your day may be broken into small intervals of time. Can you develop a list of small tasks or task that can be broken into smaller pieces that you can pick up and put down quickly so that you can be efficient with the time you have?

5. Do you take time at the end of each day to review how you have spent you energy, keeping a daily record of where you are spending your time can help you monitor how well you are balancing your commitments?

6. The measures of success may differ depending on who you ask. An informal survey of what senior faculty believe identifies a successful new faculty member at a research university resulted in the following:

   • Spent more time on research and scholarly writing.

   • Regularly sought advice from colleagues on research and teaching.

   • Integrated their research into their teaching.

   • Minimized the time they spend preparing for class.

   • Incorporated active learning techniques in their teaching.

7. What are the profiles of successful faculty at your institution?

8. How are you balancing your professional responsibilities with your personal life?

9. Do you view professional and family roles as hierarchical (you must choose to prioritize either work or family) or balanced (you can have equal weight to different roles in your life). How does your spouse/partner view this? How do your colleagues view this?

3.6.4   MENTORING CONVERSATION: DON'T TRY TO DO IT ALL AT ONCE

You can't be everybody's everything: the perfect partner to your significant other; the perfect parent/child/sibling to your family; the perfect member of your community; the perfect teacher to your students; the perfect researcher to your colleagues; the perfect committee member for your institution….. the list is just too long.

I suggest that you slow down and do some self-examination. First, it may be you who has the higher than realistic expectations of yourself. You may be pushing for more than is necessary. Or, it may be external and still unrealistic. Even if others in your life think your should be everybody's everything, it does not mean that they are right.

A friend and respected senior colleague of mine shared her philosophy with me years ago. I had asked her how she had managed to do it all. She had been a department chair (several times), associate dean, well respected researcher in her field, mother, and wife of another faculty member. Looking at all she had accomplished over the years she seemed to be superhuman. What she told me was that she takes turns. At some points in her career, her research came first, and that is what she concentrated on, giving it priority over everything else. At other points, it was her family. And at other points, it was other things. She pointed out that it would be impossible to do it all at once. Even thought this may have been her desire at times, she forced herself to make choices and set priorities. Over the course of a career, these priorities shifted from time to time enabling her to do all the things she wanted to do, but just not simultaneously!

I have had some practice at acting on her advice, but I will admit that it can be hard. It is tough to extract yourself from past engagements - first you have to let go, then you have to make it clear to others that there will be a change, then you have to have an exit plan, and finally, you have to follow your plan and make it happen. It is easy to slip up and fall back into doing the job again. After all, you may be highly accomplished in the area you are trying to let go of and the "best person for the job" but remember that no one else will be able to fill your role unless you let them. You have to leave a vacuum, and, as the saying goes, "nature abhors a vacuum." The role will be filled, and the job will get done. It may get done differently than how you would have chosen to do it, but it's likely that those differences are not really important.

3.6.5   MENTORING ACTIVITY: MAPPING YOUR BALANCE BETWEEN TEACHING, RESEARCH AND SERVICE

Step 1: Identify the categories in the table[11] below that apply to your position at your institution.

Categories Years	Years 1 and 2	Years 3 and 4	Years 5 and Beyond
Teaching, Advising, and/or Mentoring			
Scholarly Pursuits, Research, and/or Clinical Duties			
Administrative Duties, Departmental Service, Institutional Service, Professional Service, Extension, and/or Outreach			

primarily teaching   half extension   shifting over time   primarily research   evenly distributed

[11] Adapted from: **Making the Right Moves: A Practical Guide to Scientific Management for Postdocs and New Faculty,** Second Edition, Burroughs Wellcome Fund, Howard Hughes Medical Institute, 2006.

Step 2: Construct a diagram like one of the examples presented that illustrates the relative balance of time an energy you should be expending in each of these areas over time. Consider whether or not the relative balance will shift over time. Several different examples are provided for reference.

Step 3: Draft a series of major goals for each category and assign them to the three time frames listed.

Step 4: Consult with your mentor(s) about your map and ask for their suggestions on how it could be modified.

Step 5: Show your map to your department chair and/or review committee and ask how their expectations might differ from yours.

Categories	Years 1 and 2	Years 3 and 4	Years 5 and Beyond
Teaching, Advising, and/or Mentoring			
Scholarly Pursuits, Research, and/or Clinical Duties			
Administrative Duties Departmental Service, Institutional Service, Extension, and/or Outreach			

BIBLIOGRAPHY

ESTABLISHING EXPERTISE

S.A. Ambrose, K.L. Dunkle, B.B. Lazarus, I. Nair and D.A. Harkus, **Journeys of Women in Science and Engineering: No Universal Constants**, Temple University Press, Philadelphia, PA, 1997.

K. Barker, **At the Helm: A Laboratory Navigator,** Cold Spring Harbor Laboratory Press, Cold Spring Harbor, NY, 2002.

K. Barker, **At the Bench: A Laboratory Navigator,** Updated Edition, Cold Spring Harbor Laboratory Press, Cold Spring Harbor, NY, 2006.

Benefits and Challenges of Diversity, Women in Science & Engineering Leadership Institute, University of Wisconsin-Madison, `http://wiseli.engr.wisc.edu/` (Accessed 11/11/09).

L. H. Collings, J. C. Chrisler, and K. Quina editors, **Career Strategies for Women in Academe: Arming Athena**, Sage Publications, Thousand Oaks, CA, 1998.

C. I. Davidson and S. A. Ambrose, **The New Professor's Handbook: A Guide to Teaching and Research in Engineering and Science**, Anker Publishing Company, 1994.

P. J. Feibelman, **A Ph.D. Is Not Enough: A Guide to Survival in Science**, Addison-Wesley Publishing Company, 1993.

Making the Right Moves: A Practical Guide to Scientific Management for Postdocs and New Faculty, Second Edition, Burroughs Wellcome Fund, Howard Hughes Medical Institute, 2006.

R. M. Reis, **Tomorrow's Professor: Preparing for Academic Careers in Science and Engineering**, IEEE Press, 1997.

Tomorrow's Professor eMail Newsletter, sponsored by Stanford Center for Teaching and Learning, subscribe through `http://cgi.stanford.edu/~dept-ctl/cgi-bin/tomprof/postings.php` (Accessed 11/11/09)

Training Scientists to Make the Right Moves: A Practical Guide to Developing Programs in Scientific Management, Burroughs Wellcome Fund, Howard Hughes Medical Institute, 2006.

TEACHING:

K. Bain, **What the Best College Teachers Do**, Harvard University Press, Cambridge, MA, 2004.

S.A. Baiocco and J. N DeWaters, **Successful College Teaching: Problem-Solving Strategies of Distinguished Professors**, Allyn and Bacob, 1998.

Committee on Science, Engineering and Public Policy, **Reshaping the Graduate Education of Scientists and Engineers**, National Academy of Sciences/National Academy of Engineering/Institute of Medicine, National Academy Press, 1995.

B. Gross Davis, **Tools for Teaching**, Second Edition, Jossey-Bass Publishers, 2009.

J. Handelsman, S. Miller, C. Pfund, **Scientific Teaching**, The Wisconsin Program for Scientific Teaching, W.H. Freeman and Company, New York, NY, 2007.

R. Magnan, **147 Practical Tips for Teaching Professors**, Atwood Publishing, 1990.

J. K. Reklaitis, **Small Groups for Students in Engineering**, Purdue University , Women in Engineering Program, West Lafayette, IN, 1999.

S.D. Sheppard, K. Macatangay, A. Colby, W.M. Sullivan, **Educating Engineers: Designing for the Future of the Field**, The Carnegie Foundation for the Advancement of Teaching, Stanford, CA, 2009.

P.C. Wankat and F. S. Oreovicz, **Teaching Engineering**, McGraw Hill, 1993.

ADVISING AND MENTORING STUDENTS:

A.W. Astin, **What Matters in College: Four Critical Years Revisited**, Jossey-Bass Publishers, San Francisco, CA, 1993.

AWIS, **Mentoring Means Future Scientists: A Guide to Developing Mentoring Programs Based on the AWIS Mentoring Project**, The Association for Women in Science, Washington, D.C., 1993.

Committee on Science Engineering and Public Policy, **Advisor Teacher Role Model Friend: On Being a Mentor to Students in Science and Engineering**, National Academy of Sciences/National Academy of Engineering/Institute of Medicine, National Academy Press, 1997.

T. Goodwin and K.E. Hoagland, **How To Get Started in Research,** Council on Undergraduate Research, Washington, DC, 1999.

B.B. Lazarus, L.M. Ritter, and S.A. Ambrose, **The Woman's Guide to Navigating the Ph.D. in Engineering and Science**, IEEE Press, New York, NY, 2001. DOI: 10.1109/9780470546673

J. Nakamura, D.J. Shernoff, C.H. Hooker, **Good Mentoring: Fostering Excellent Practice in Higher Education**, John Wiley & Sons Inc., San Francisco, CA, 2009.

RESEARCH AND SCHOLARLY ACTIVITIES:

Committee on Science, Engineering and Public Policy, **On Being a Scientist: Responsible Conduct in Research**, National Academy of Sciences/National Academy of Engineering/Institute of Medicine, National Academy Press, Washington, D.C., Third Edition, 2009.

GRANTS AND FUNDING:

ACE/Kellog Faculty Workplance Project, **Survival/Success Guide for Knox Faculty**, 4th edition, Knox College, Galesburg, Illinois, 2004, `http://deptorg.knox.edu.facdev/guide/` (Accessed 11/11/09).

InfoEd Internations Inc., **SPIN Search**, `http://www.infoed.org/new_spin/spin.asp` (Accessed 11/11/09).

ProQuest LLC, **Community of Science (COS) Funding Opportunities**, `http://fundingopps.cos.com/` (Accessed 11/11/09).

University of Illinois at Urbana-Champaign, **IRIS: Illinois Research Information Service**, `http://www.library.illinois.edu/` (Accessed 11/11/09).

U.S. Department of Health and Human Services, **GRANTS.GOV**, `http://www.grants.gov/` (Accessed 11/11/09).

THE BALANCING ACT:

S.D. Friedman, **Total Leadership: Be a Better Leader, Have a Richer Life**, Harvard Business School Publishing, Boston, MA, 2008.

I have been fortunate to have had exceptionally wonderful and generous mentors over the years. I have also found that mentoring comes in many forms and under interesting disguises. Some of my mentors are probably not even aware that they fill this role for me, but others are involved with me in very structured relationships. The best advice that I can give is to take advantage of any and all mentoring that comes your way. You don't have to choose to follow the path that a mentor is advocating, but you should listen with an open mind and gather as much information as possible before making a decision.

CHAPTER 4

Developing Networks, Relationships, and Mentoring Activities

The academic profession is a "colleague system." Your relationships partly control and shape your place within your profession and your field. You can negotiate your way through such a system by establishing a variety of connections and relationships on multiple levels.

One of the types of relationship you will want to develop is often termed a mentoring relationship. This is actually a broad category and can be thought of in classical terms, where your mentor is one key individual who will apprentice you in your field, or in more contemporary terms, where mentoring occurs on multiple levels with multiple individuals and incorporates peers, professional networks, and colleagues as well as the classical mentors. Having mentors in both categories is valuable and will contribute to a successful career.

Research by Prof. Monica Higgins of Harvard's Graduate School of Education and Prof. David Thomas of Harvard's School of Business supports the idea that broader models of mentoring are effective when longitudinal data is taken into account. Their study compared the impact of individual mentors with "constellations" of supporting individuals in a variety of mentor-related roles. They say that, "While the quality of one's primary developer affects short-term career outcomes, it is the composition and quality of an individual's entire constellation of developmental relationships that account for long-run protégé career outcomes."[1] In other words, having many mentors will provide you with the best opportunities for success in over the course of your career.

The key to getting good mentoring is being receptive - to new relationships, criticism, advice, and support. Seek advice and develop relationships with the people you are connected to through your networks and collegial relationships. If you are not connected to the right people, then set up your own networks that will ensure your success.

One of the biggest pitfalls in mentoring and all collegial relationships is the disparity in power that exists between individuals because of the structure of your institution and your field. Here again, information is the key to navigating what can sometimes be sticky issues when an imbalance in power enters into the equation.

---

[1]M.C. Higgins and D.A. Thomas, "Constellations and careers: Toward understanding the effects of multiple developmental relationships," **Journal of Organizational Behavior,** 22, p. 223-247, 2001.

4.1 GETTING THE MENTORING YOU NEED

4.1.1 OVERARCHING QUESTIONS TO CONSIDER

- What role models have you identified in your field?

- How do you envision the role that your mentor(s) will play in you career development?

- Can you find the mentoring that you need at your institution?

4.1.2 MENTORING CONVERSATION: ON FINDING THE ROLE MODELS AND MENTORS YOU NEED

As a new assistant professor, I found one of my most helpful mentors through a chance conversation. He turned out to be one of the few senior male mentors and advisors who truly treated me as a colleague and for that alone he is irreplaceable. Aside from helping me to develop my research program and learn how to secure funding, he showed me how to blend a high quality research program with a true commitment to education. His mentorship was unexpected, especially given that he was from an entirely different field in a different college at my institution. But, if you are open to and seek out these relationships, they can have a lasting positive imprint on your career.

Beyond this individual and several other colleagues within my department who provided me with invaluable mentoring in the first few years of my faculty career, I also joined and created several peer mentoring groups over time. These were often rather informal groups but what made them each work was that, at that particular moment in time, we were all interested or in need of the same things and could provide a space for conversation, advice, and support to each other. One example of a group I joined as an assistant professor was sponsored by the university for faculty professional development, called Creating a Collaborative Learning Environment. In this case, the group was more formal and had a staff facilitator and a syllabus that we engaged with over the course of two semesters. The group was very dynamic because the faculty came from a variety of disciplines across campus, but we all had a common interest in exploring how people learn.

Later, I was involved in a peer mentoring group that was formed at the grass roots level. Six faculty women in our college had their first child within two years of each other, and we established our own network of support – a peer mentoring group. In additional to standard "mom talk", we discussed other topics like the challenges of balancing a faculty career and children, the pros and cons of "stopping the clock" for tenure, strategies for arranging child care to fit our own individual needs, and how to effectively juggle travel obligations with pregnancy and/or kids in tow. In a few short years, that group grew to more than 20 women faculty members, all of whom were expecting or have young children. We meet regularly for lunch and have an active listserve for advice exchange via email. This group has been a huge source of support and friendship over the years.

Even though I have highlighted a few particular examples above, they are really just a few of many. Searching for a single mentor who will provide you with advice on everything you want to be in life, both professionally and personally, is almost always a fruitless quest. The more likely reality is that you will need many role models, mentors, and peer supporters, so that a patchwork quilt can be pieced together that is it styled after your own individual nature and your goals.

4.1.3   DETAILED QUESTIONS AND SUGGESTIONS TO CONSIDER

1. What do you expect of a mentor? Do you expect guide, advocate, and/or friend to provide:

   - positive and constructive feedback?
   - assistance in developing networks?
   - guidance in setting goals?
   - understanding and empathy?
   - encouragement and nurturing?
   - socialization into your discipline?

2. In building a set of mentoring relationships, you will want to consider the range of your needs and how each individual mentor might help you with different aspects of your life and career. There are several topics to consider within mentoring relationships:

   - Do you feel comfortable asking for advice and accepting criticism?
   - Can your mentor meet with you frequently enough for your needs?
   - How formal do you want the relationship to be?
   - Can you develop a productive and non-threatening relationship with your mentor?
   - Should you share everything with your mentors or be selective about what you discuss with each individual?
   - Is the individual familiar with your academic position and your institution?

3. What types of issues do you want to discuss with your mentor? Some possibilities include the following:

   - teaching/research/service
   - balancing family and career
   - managing your time
   - setting goals
   - preparing your tenure packet
   - handling conflict
   - dealing with department politics
   - understanding the culture of your discipline

- knowing when and how to say "No"

4. It will not be desirable to get all of your mentoring and advice from one individual. How will you go about developing a set of mentoring relationships to meet your needs? Consider approaching some of the following people:

- Colleagues in your department
- Colleagues outside your department and at other institutions
- Formal mentoring programs offered by your institution or professional society
- Other junior faculty who can provide peer mentoring
- Individuals in the community not associated with your university

5. Have you found ways to be openly appreciative of the mentoring you receive? You might consider:

- thanking the person (verbally of in writing) even if it has been years since they mentored you.
- acknowledging their mentoring efforts publicly.
- nominating them for a mentoring award.

6. Do you have both positive and negative role models? How can you use the positive and negative models to motivate you in your professional development?

4.1.4   MENTORING ACTIVITY: STARTING YOUR OWN PEER MENTORING GROUP

Step 1: Identify a common topic of interest for the group, e.g., proposal writing, teaching in your disciplinary area, parenting while professing, etc.

Step 2: Identify a few colleagues who you would like to invite to join you in the group. Have a conversation with each of them about their interest in meeting regularly to discuss the topic. Identify the best venue for the meetings and timing for the meetings. Take into account that some members may have other obligations that prevent them from meeting at certain times or on certain days of the week.

Step 3: Set up an email group or listserve with the initial members and send out a formal announcement, e.g., "Thank you for agreeing to join me in our new peer mentoring group on TOPIC. I have reserved ROOM/BUILDING for our first meeting on DATE/TIME. At this initial meeting I propose that we develop an agenda for our group and plans for our future meetings over the coming academic year."

Step 4: Develop consensus within the group about the formality of meetings, frequency of meetings, optimal size of the group, and responsibilities of the group members.

Step 5: Grow the group to a sustainable size. This can be accomplished through the networks of the initial group members or speaking with a campus administrator in a closely affiliated office or program about other individuals they may know of who would be interested in the group.

Step 6: As the "convener" of the group, you will be responsible for sending out reminders for meetings and keeping the momentum of the group going. It is good practice to rotate the "convener" responsibility to a new individual for a group that meets for more than one year.

4.2 BECOMING A COLLEAGUE?

4.2.1 OVERARCHING QUESTIONS TO CONSIDER

- How will you separate yourself from you former role as a student or post doc in the eyes of your existing network of mentors and colleagues?

- What type of colleague will you be to the faculty and staff in your department?

- Can you see a path from junior colleague to peer of the senior faculty?

4.2.2 MENTORING CONVERSATION: ON THE ROLE MENTORS CAN PLAY IN OVERCOMING OBSTACLES

Looking back, I think one of the low points of my academic career was at my first funding agency meeting where all of the principal investigators currently funded by a program gathered to talk about their research. It is a fairly competitive environment because you are essentially expected to defend your work in front of your rivals and the program manager while your funding is at stake.

Being new to the ranks of principal investigator, I had not met any of the other attendees in the room (aside through their authorship of journal articles I had read). Being the only woman in the room – a phenomenon that still occurs in engineering – I felt like even more of an outsider. The feeling was magnified when, as everyone took their seats, I discovered that a 360° empty chair buffer had been created between me and everyone else in the room. That day was one of the longest and loneliest of my life. It made me seriously question whether or not I really wanted this job.

This is the sort of situation where a mentor can play a pivotal role in a junior colleague's career. I mentioned what had happened at the meeting to one of my senior colleagues in the department when I returned. He asked to see the list of attendees, and then he identified a couple of conferences that we could both attend in the coming year. Introductions to key principal investigators within this group at conferences over the following 12 months made the next year's meeting a pleasant experience. It is amazing how a simple connection like a chat between sessions or being included in a dinner group can build one's network. By the time year two rolled around, I could walk into the room, spot familiar faces, and initiate a conversation by simply saying "Hi, it's nice to see you again."

4.2.3   DETAILED QUESTIONS AND SUGGESTIONS TO CONSIDER

1. Teaching, research and service are the traditional triumvirate in faculty evaluation, but you should be aware that collegiality may play an unwritten role in the decision. You are in the process of developing life-long relationships with your current colleagues.

   - What relationships are you establishing with other faculty, student, and staff?
   - Do you have positive and respectful relationships with other faculty and staff members in your department?
   - Are you considered to be a "good citizen" of your department or institution?
   - Whom can you assist? Who can assist you?
   - How can you help these relationships develop positively?

2. The staff members of an institution – clerical support, administrative assistants, computer support, grant management officers, instructors, etc. – are often the key to actually getting things done. Who are the individuals critical to your success? Work towards establishing good relationships with these people.

3. Are you developing collaborations with other faculty who can help you advance professionally? How is collaboration valued or discounted within your department? Can you clearly define the contributions that you bring to the collaboration?

4. In what ways can you use the university organizations or programs that are designed to aid your professional and/or personal development?

5. Consider how much information about your personal life you want to share with your colleagues, mentors, and students, including details about your:

   - recreational activities,
   - health issues,
   - family life,
   - partner/spouse,
   - sexual preference, and
   - lifestyle choices.

4.3   BUILDING NETWORKS

4.3.1   OVERARCHING QUESTIONS TO CONSIDER

- What strategies can you use for creating and broadening your professional network?
- Are there steps you can take to actively maintain your network?

4.3.2   MENTORING CONVERSATION: ON STRATEGIES FOR DEVELOPING YOUR NETWORKS

In most fields, developing professional networks and collaborations are essential for success. (You may need these networks to be local, regional, national or international depending on your field and institutions. For some, developing an international scholarly reputation, or some such equivalent, is part of the criteria used for tenure decisions at your institution.) Professional networks come in many varied forms, but all involve developing professional relationships with other faculty and researchers in your field or related fields. Although networks take some work to maintain, they can be grown in a very natural fashion by putting yourself in the right place and taking advantage of opportunities.

One key way to build a network within your discipline is to be involved with a professional society in some way. You should avoid the roles that are a time sink especially if they do not put you in a position to make new contacts within your field. However, there are good choices, such as co-organizing a symposium for an upcoming meeting, or a workshop on your specialty area. The process of contacting invited speakers will give you the opportunity to connect with key senior faculty around the world and the process of chairing or hosting will provide you with visibility.

I have organized conference symposia, co-chaired sessions for professional meetings, and coordinated workshops at various stages in my career. Each was a valuable and fruitful experience. In every case, I have come away with new contacts and new professional opportunities that made the time investment worthwhile. Often, they have led to invitations to speak in other venues, which increased my exposure to people in my discipline even further.

In addition to the professional networks, research collaboration with colleagues is becoming standard practice in many fields. In some cases, collaboration is born out of necessity because that is where the funding is available, but in other cases, it is the interesting questions that lie in the spaces between traditional disciplines that encourage the growth of collaborative partnerships. Steve Ackerman posits that, "Bringing together people of varied backgrounds and skill sets to inspire alternative ways of thinking provides new approaches to solving complex problems."[2]

Collaboration can be amazingly fruitful, energizing, and intellectually stimulating when it works, but it can also be a nightmare that makes you dread every meeting with your collaborators and make you wish you had never started the project. My personal rule is that I will only collaborate with people that like. We have to get along and respect each other over the duration of the collaboration, and it is likely that there will be challenges to face and problems that will test the relationship. This is why I feel it is critical for the personalities of the collaborators to be compatible at a minimum, and best if you can enjoy a friendship and camaraderie as well as complementary expertise. Once a good collaborative relationship is established, then I have found it often to be the case that we continue to look for excuses to work together even after the original project is completed. It is these relationships that make it all worthwhile.

---

[2] S.A. Ackerman, "Developing Positive Team Collaborations," Bulletin of the American Meteorological Society, p. 627-629, May 2007.

4.3.3   DETAILED QUESTIONS AND SUGGESTIONS TO CONSIDER

1. You are probably participating in a variety of networks already. What other groups might you join to expand your networks? Pursue opportunities to meet peers in your department and college as well as beyond your institution. Don't forget to maintain relationships with former co-workers, advisors, teachers. You should also consider new options for network building:

   - Committee service
   - Internet groups
   - Professional organizations
   - Social and activity-based organizations
   - Faculty governance and other organizations
   - Community groups

2. How can you get to know people in your field locally, regionally, and internationally? In what ways can you use the internet to meet people and market yourself? Are you making connections and building relationships with faculty inside and outside your discipline? On and off campus?

3. What are you looking for in your network connections? How can you introduce yourself to and build positive relationships with people in these networks? What do you have to offer them? What do they have to offer you?

4. What steps are you taking to meet senior people in your field? How are you informing them of your research? Consider:

   - Inviting a senior person to give a seminar in your department
   - Organizing the department seminar for a semester or year
   - Making sure that your desire to give seminars at other institutions in known throughout your network
   - Meeting with all seminar speakers that visit your department so that you can discuss mutual research interests
   - Giving talks at meetings and conferences
   - Serving on review panels
   - Sending out reprints to people in your field, particularly to those whose work you reference

5. Networking within your disciplinary research area is essential. Much of this networking is done at conferences where there is often a key person in your field that you would like to meet. Have you considered contacting that person before the conference to introduce yourself and set up a brief meeting with them during the conference?

6. Developing your network is the first step, but maintaining that network is also vital. How do you maintain your networks? Consider the importance of:

   - performing some type of follow-up contact after a first meeting,
   - attending the same conferences every year,
   - sending preprints or reprints of your work to people in your research network, and
   - sending articles of interest or information about opportunities to people in your broader network.

7. How are you projecting yourself electronically? Part of your image in the professional world is established by your internet presence. In addition to keeping your faculty webpage attractive and up to date, ensure that the image you project is professional even on your FaceBook page. The internet will ultimately be a resource for people trying to learn about you when considering whether or not they want you as a graduate advisor, a collaboration, or a letter of recommendation.

8. Networking is an essential but time-consuming activity. Think carefully about how much time you need/want to devote to networking activities. How have you balanced the time you spend on networking with your other responsibilities?

4.3.4   MENTORING ACTIVITY: GROWING YOUR NETWORK

Step 1: Identify your goals[3]:

   - Do you want to improve your network to facilitate your research?
   - Do you want to develop your reputation outside your institution?
   - Other purposes?

Step 2: Build a list of contacts:

   - Identify relevant people. Who do you already know? Who do you need to know?
   - If you don't know individual names, identify the types of people you need in your network and then seek out individuals who are that type.

Step 3: Develop strategies to court these people individually:

   - Talk to your colleagues and mentors about who they know.
   - Identify professional organizations where you might meet people important for your network.
   - Work to identify a way to meet each person face-to-face.

---

[3]These steps are adapted from: D.R. Faught, "Developing Your Professional Network," **STQE**, p. 72-73, January/February 2001, www.stqemagazine.com (Accessed 11/11/09).

- Reciprocity is important. Don't always ask for something every time you interact with a person. Don't ask for something at your first contact with someone you have just met. Try to figure out a way for you to give something. This is why it is important to build your network before you need it.

Step 4: Maintain your network. What can you do to follow up occasionally with people in your network? Schedule time each week to tend your network.

Example:

Goal: Expand research connections to improve visibility in my field.		
Current and Potential Contacts	**Courting Strategy**	**Maintenance Plan**
A.B. Smith, Ph.D. advisor	Current contact	Send email update every 3 months
C.D. Jones, former student colleague from Smith group	Current contact	Invite her to give a talk at my institution, and ask her to reciprocate at her new institution
E.F. Big, senior researcher in my filed	Introduce myself at next Big Research Conference after his talk; ask my former advisor to invite us both to lunch	Send him a copy of my reprints as my publications occur; plan to attend Big Research Conference annual and go to his talks

Goal:		
Current and Potential Contacts	**Courting Strategy**	**Maintenance Plan**

Goal:		
Current and Potential Contacts	**Courting Strategy**	**Maintenance Plan**

Goal:		
Current and Potential Contacts	**Courting Strategy**	**Maintenance Plan**

4.4   DEVELOPING A REPUTATION-MAKE YOURSELF VISIBLE!

4.4.1   OVERARCHING QUESTIONS TO CONSIDER

- In what ways can you be strategic about making yourself visible to your colleagues?

- Have you identified strategies that you are comfortable pursuing?

- Can you work with your mentors to identify ways to improve your visibility in a positive way?

4.4.2   MENTORING CONVERSATION: ON POSITIVE INTERACTIONS WITH YOUR COLLEAGUES

I don't subscribe to the "be seen but not heard" philosophy for assistant professors, but I do advise junior faculty members to exercise some caution when jumping into the fray. You need to strike a balance between gaining visibility while not risking your professional reputation. Because of this, it is important to try to understand some of the history of a group – like your department – before engaging in a faculty meeting discussion for instance. There could be hidden land mines. Also, at professional meetings – especially during the Q&A period or open discussion – there are likely to be certain unwritten rules of conduct and even "factions" within the discipline. Marching in without knowing some of this background can get you into trouble or unintentionally identify you as taking a side in a long standing argument.

There are a few strategies that you can employ to avoid finding yourself at the center of an argument or looking naïve in front of your colleagues. The most severe or hypercautious strategy would be the "be seen but not heard" approach mentioned above. This is not necessarily the best strategy, however, because it can make you invisible or, worse, seen as someone not willing to engage. Another option is to tread lightly into the conversation with clarifying questions. This can be a particularly useful strategy in faculty meetings because your questions can help you to glean information about the history, the people involved, and why things have been done in certain ways in the past.

The best strategy is to identify trusted mentors. These people may even be junior colleagues who are just a year or two ahead of you. If they have been keeping their eyes open and doing their homework, they should have developed some insights into the social dynamics of the people in your department and your discipline. Your senior colleagues know all of the history and the players involved, so a trusted senior mentor can help you to identify the best strategy for making your point, bringing up a new proposal, and gaining visibility in your field.

As I write this, I am reflecting on a scene at a professional meeting I attended recently that may serve as a cautionary tale. The meeting organizer had set aside time for discussion of future research directions and asked one of the attendees to facilitate the discussion. A few minutes into the discussion, one of the other attendees hopped up, stood next to the facilitator, and proposed that, instead of the brainstorming that we were currently doing, we should try to phrase everything in terms of a dilemma. He was very persistent, rephrasing what people were saying and letting them know that the points they were trying to make were not dilemmas. Unfortunately, this did not seem to be an effective strategy with the group. A while into this, I turned to a colleague and whispered "Who's dilemma guy?" He responded "He's an assistant professor making a big mistake." At the end of the hour, the coordinator finally spoke up and said that this really was not what he was hoping for in our discussion and gave us more detailed instructions for what he would like to see during the discussion time on the second day of the meeting. The assistant professor had made a big tactical error in his approach, and I felt badly for him. He had made himself visible, but he did not built the sort of image for himself that is considered to be especially positive in our field. It was unfortunate that the facilitator did not redirect his enthusiasm and that the meeting organizer had not been more clear about his goals up front.

4.4.3 DETAILED QUESTIONS AND SUGGESTIONS TO CONSIDER

1. Getting involved in conferences and meetings is important and will help you to mature in your discipline. Which people should you meet at these conferences? How do you initiate and maintain these relationships?

2. How often have you published your work in appropriate peer-reviewed journals and conference proceedings?

3. How often do you write about your work and submit publications and grant applications? Are you satisfied with this rate of publication and grant consideration? How might you increase your publication and grant proposal acceptance rates?

4. Do you regularly take time to have discussions with and seek advice from your colleagues?

5. Does your personal communication style fit your field? What are the implications of gender differences in language use? Consider these general suggestions:

   - Express certainty when you give and opinion or state a fact.
   - Phrase ideas in a win-win format by explaining how it will benefit you and the other person.
   - Don't take it personally when someone disagrees with your idea.

6. What kinds of recognition and reward systems exist at your institution? Are you utilizing them for yourself and for others? Are some rewards/recognitions missing, and can you find ways to create them for yourself and others?

7. Have you taken time think about how you would explain aspects of your scholarship to non-experts? Does your institution offer media training? Consider the following tips for honing your message: [4]

[4] R. Hayes, D. Grossman, **A Scientist's Guide to Talking with the Media: Practical Advice from the Union of Concerned Scientists**, Rutgers University Press, New Brunswick, NJ, 2006.

- Keep it simple but don't make it simplistic.
- Avoid the jargon of your discipline.
- Think in advance about how you can. get across your key message in a sound bite.

4.4.4   MENTORING CONVERSATION: RESOLVING CONFLICTS

What do you do when you find yourself in a heated argument or ongoing conflict with a student, colleague, or a department chair? These sorts of situations can be tense, draining, and potentially career damaging. Certainly, one goal might be to avoid conflict or minimize it when it first occurs, but occasionally the development of a conflict may be outside of your control. For example, a person maybe harboring feelings of being wronged or slighted before you are even aware of an issue. In other instances, you may choose to enter into a conflict because you feel that the battle is a worthy cause or that you have no other choice if you want to remain true to yourself. Unfortunately, conflicts that carry on for an extended time can be very damaging. Without resolution you, and probably everyone involved, are paying a price with compounding interest.

A number of my faculty friends have, over the course of their careers, run into untenable situations and personnel-oriented problems. These can be very disruptive and upsetting when they happen, and if the situation involves another faculty member, there is risk that it may influence your tenure decision or be a lifelong problem until one of the two of you retires. The best advice I can give is to seek help from a confidential and experienced source. Many institutions have ombudspersons or an office of conflict resolution. In some cases, you might also consider seeking advice from a professional counselor or lawyer outside the institution as well. It may seem to you as though you have few viable options, but there are often creative solutions and positive outcomes that can be pursued.

In addition to seeking advice from a third party to help you negotiate the situation or involving them as a facilitator to bring about resolution, there are a number of things you can do early on to avoid escalation of the issue into an outright conflict. One strategy is to try to see all sides of the issue and understand the background they may have come from before you even arrived on the scene. Be open to conversation and listen to the perspectives of others. Avoid becoming entrenched in your opinion. If you think there is only one right answer, the issue is bound to escalate. In most cases, if you can be flexible, then there are creative solutions that you can seek out. Sometimes a bit of self-discovery is required in this regard. Reflect on what is really behind your thinking on the issue and identify your needs versus your preferences. Finally, when a resolution is developed, make sure everyone is clear on its meaning and what actions are to be taken as a result.

4.5   MAKING PRESENTATIONS

4.5.1   OVERARCHING QUESTIONS TO CONSIDER

- How will you go about developing the presentation skills that you need to be successful in your career?

- Can you identify strategies for improving your presentation skills that will be effective for you?

4.5.2   MENTORING CONVERSATION: ON BECOMING A BETTER PRESENTER, A TASK IN CONTINUOUS IMPROVEMENT

I am multitasking as I write this piece. I am at a technical meeting listening to a talk as I am making notes on sections that I would like to add to this book. Maybe my full attention should be on the speaker, but it is next to impossible for me to do so. It is the middle of the afternoon and the speaker is awful. I've seen worse, but he is really at an extreme case on the scale. Sadly, he is a bit better than the previous speaker, but I won't even dwell on that.

What occurs to me as I look around the room is that, because his presentation skills are so bad, very few people are actually listening to him. He's missing - and we're missing – an opportunity. He would like to communicate to us about his work, tell us why it is exciting, and maybe even prompt a few of us to interact or even collaborate with him in the future. This loss is not because what he is talking about isn't intrinsically interesting (I made myself listen for a few minutes and there was some good stuff buried in the talk). The loss is occurring because his presentation style makes it very difficult to listen. He has not looked at the audience once, he's mumbling, his voice is flat, and his explanations are convoluted.

What's the lesson? Do whatever it takes to figure out how not be this guy! Observe and emulate good speakers. Ask someone whose style you admire to critique a practice talk. You might even consider utilizing the services for improving communications skills intended for students at your institution if there are none specifically targeted to faculty. Because being able to communicate your work in a clear and compelling manner is critical, you need to treat every talk you give to your colleagues as an opportunity to advance your career.

I will confess that I have not always been a very good public speaker myself. In fact, as an adolescent I was painfully shy. But I made a decision upon embarking to high school that I was going to change. It was exceptionally uncomfortable at first, but I went out of my way to put myself in more social situations and even joined the debate team so that I would have to speak in front of people. Things did not change overnight, but many years later, with lots of practice, most people consider me to be a good speaker. However, I still consider my skills to be a work in progress, and I am constantly seeking to make improvements. One technique I use is that after I see a really good talk, I take a few minutes to reflect on and sometimes even write down what made the talk good and the techniques or ideas that I could emulate or adopt. I also regularly ask for feedback – from students or a close colleague in the audience.

Undertaking an effort to improve your communication skills is worthwhile. The next day in this same meeting I mentioned above, I saw another speaker who showed how much a person can improve in just a year. I remember his talk from last year. The content was good and well organized, but his presentation skills were a bit rough, and he was very nervous. This year he gave one of the best talks of the week. He was confident, well spoken, engaged the audience, and showed his enthusiasm for his work. It was really exciting to see the transformation. I imagine, he put a lot of work into making it happen, and it had paid off.

4.5.3   DETAILED QUESTIONS AND SUGGESTIONS TO CONSIDER

1. What experience do you have in presenting and defending your ideas in friendly settings (a journal club or other less formal setting) and in more formal settings (departmental seminars, national conferences)? How can you use your experiences to further your goals?

2. Have you presented and defended your work at poster sessions, conference talks, seminars, and other events at departmental, university, community, national, and international conferences? Have you asked others to critique your presentations and provide additional opportunities to develop your presentation skills?

3. In preparing to give conference and seminar presentations, it is important to keep a number of things in mind:

   - Keep your presentation focused on the key issues of your research.

   - Give credit to others - colleagues, collaborators, students.

   - Identify the motivation of your research.

   - Review work on the topic reported in the literature and cite the work of others.

   - Clearly, identify your unique contributions.

   - Delineate the findings and implications of your research to your field.

   - Spend time refining your presentation material so that it is concise and easily viewed.

   - Have a contingency plan if you are relying on technology in your presentation.

   - Rehearse your talk.

   - Stay within the time limitations and leave time for questions.

   - Speak clearly and be sure that you project loud enough so that people sitting in the back of the room can hear you.

   - Involve your audience (pose rhetorical questions; build in interactive workshop exercises).

   - Ask your peers for constructive feedback after your presentation so that you can improve your delivery.

4. Have you considered how to ensure that your actions and the manner in which you present yourself and your ideas will be reviewed? Do you present a positive and ethical image?

4.6   PERSONAL REFERENCES

4.6.1   OVERARCHING QUESTIONS TO CONSIDER

- Do you know how letter writers will be chosen when it comes time for tenure?

- How will you structure an effective letter of recommendation for one of your students?

4.6.2   MENTORING CONVERSATION: ON BEING A GOOD LETTER WRITER

It is letter writing season again. Undergraduates from courses and my research group are applying to graduate school and students of every ilk are looking for scholarship and fellowship funding. I've been averaging about two letters of recommendation a day for the last few weeks. It is a part of the job of a faculty member that does not get mentioned much but can take up considerable time depending on the number of students you interact with regularly. Now that most of these are done electronically, the burden seems a bit lower. I don't have to worry about whether I remembered to sign over the sealed envelope flap!

Over the years, I have developed some strategies to make my letter writing more effective. When a good student who I am well acquainted with asks me for a letter of recommendation, I ask them to provide me with a copy of their resume, a draft of their essay, a copy of the instructions for recommenders, and a short list of any key points they want me to try to address in their letter. A quick review of these items makes my job much easier because I have all of the facts at hand and a idea of the scope that the letter will cover when I sit down to write.

In most cases, a student will ask for multiple letters and maybe even letters over the course of several years. This means an initial investment of time for me, but I can quickly modify the initial letter to meet future needs. However, the initial investment in a good letter of recommendation is substantial (at least a half an hour) because it must touch on a number of topics. The letter itself needs to provide the context of how the student compares to others in the same cohort. It is also critical to provide some level of detail regarding the student's accomplishments. When possible, I try to provide a short story about a particular class or research project the student has worked on and what their contributions were. I also try to provide measurable outcomes. This can take the form of GPA or grade achieved in a course, papers written, or talks given, for instance. I also talk about how the student works with others and their leadership skills. Finally, it is critical to address all of the criteria laid out in the instructions to recommenders. Omissions may be taken as a negative reflection on the student.

Occasionally, a student who I don't think highly of will ask me for a letter of recommendation. My usual response to this is to say "I'm not sure I can write you the strongest letter. There is likely someone else who can." If they are insistent, I tell them the general flavor of what I think I can say in their letter so that they know its limitations and ask if they know someone else who would be able to write a more complete and compelling letter. Still some insist. My strategy then is to spend no more than five minutes writing a letter that states the basic facts in a single paragraph. Brevity alone sends a strong message to the reader.

Let me add a brief word here on the issue of unconscious bias and how it impacts letter writing, both as the writer and as a reader: Everyone carries unconscious biases about people that fall into different categories from their own, whether it is gender, race, age, or something else. You are not exempt from the biases even if you are in the group yourself; both men and women carry the same assumptions about gender, for instance.[5] For example, letters written for women are often shorter, do not provide solid evidence of accomplishment, and use gendered descriptors.[6] But the more you learn about how biases can affect your recommendations and your interpretation of letters written by others, the less unconscious bias will play a significant role in your behavior.[7]

4.6.3   DETAILED QUESTIONS AND SUGGESTIONS TO CONSIDER

1. Identify relationships you have developed with professionals who can write strong, positive references for your tenure file. In what ways are you maintaining these relationships? How will you establish new relationships? For example, how often do you interact with senior professionals in your field?

2. Who can provide the most appropriate reference for a given situation? Identify whom you would approach for letters of support or nomination in a variety of situations:

   • Fellowship applications

   • Teaching awards

   • Committee assignments

   • Officer positions in professional organizations

   • Tenure packet

3. Tenure packets require letters from people in your field commenting on your research ability. Have you determined whether or not you have any formal or informal input as to who is asked to write letters for you? What formal and informal criteria are used to choose these individuals?

4. Have you considered what those writing references for you will say? Are you prepared to provide information to recommenders that you contact? Are you comfortable coaching a potential recommender on what areas of your career you would like them to highlight? A letter of recommendation should include such information as:

---

[5] M. Biernat, M. Manis, T.E. Nelson, "Stereotypes and standards of judgment." **Journal of Personality and Social Psychology**, 60(4), p. 485-499, 1991.

[6] "Reviewing Applicants: Research on Bias and Assumptions." Women in Science & Engineering Leadership Institute, University of Wisconsin-Madison, `http://wiseli.engr.wisc.edu/` (Accessed 11/11/09).

[7] P.G. Devine, E.A. Plant, D.M. Amodio, E. Harmon-Jones, S.L. Vance, "The regulation of explicit and implicit race bias: The role of motivations to respond without prejudice." **Journal of Personality and Social Psychology**, 82(5), p. 835-848, 2002.

- work history,

- educational history,

- those specific skills relating directly to the position that make you stand out (e.g., specific teaching methodology, research methodology, and equipment use),

- action verbs and quantified data about your accomplishments, and

- relevant personal strengths with examples (e.g., communication skills, teamwork, and leadership abilities).

4.7 ANALYZING POWER RELATIONSHIPS

4.7.1 OVERARCHING QUESTIONS TO CONSIDER

- Who can help you to navigate your place in the hierarchy of your department and institution?

4.7.2 MENTORING CONVERSATION: ON HOW POWER RELATIONSHIPS CHANGE OVER TIME

One of the challenges of being in a faculty position is that there is very little turnover. Your faculty colleagues will be roughly the same group of people ten years from now as they are today. Over the decades, new hires will occur and retirements will happen, but the many of your colleagues will be with you for the duration. In some ways, it is like being a part of a family, crazy uncles and fun cousins included.

The relationships, however, do change over time. This often naturally occurs as a new assistant professor matures and grows in experience, morphing from a junior colleague, to a colleague, to a senior colleague over the course of their career. It may happen that the people most valuable to you as mentors when you were new are less interested as you establish your footing and develop your own identity within the department. There may be other colleagues at this point who now find you more interesting because you can be a full partner in the relationship, and they may gain as much through interactions with you as you gain from them.

There can be, however, a bit of a dark side to the changes that occur over time. A wonderful colleague of mine wrote an editorial on the "Jennifer Fever" phenomenon. In my experience, it holds more broadly than for just academic medicine. The story goes that the "Jennifers" newly entering the field are taken under the wing of senior men, while the "Janets," who are currently senior women, are ignored and marginalized. But the "Jennifers" will eventually become the "Janets" themselves. As they advance in their careers "where they have competence, experience, opinions they may wish to voice, and a legitimate claim on resources, they are abandoned."[8] The formerly-young women who believed that gender issues are no longer a problem begin to discover that they still persist but in a different form.

---

[8] M. Carnes, J. Bigby, "Jennifer fever in academic medicine." **Journal of Women's Health**, 16(3), p. 299-301, 2007.

Regardless of how it manifests itself, your place in the power relationships that you will interact with over your career will change. Knowing that these power relationships exist and understanding how they might impact you are essential to navigating your way effectively.

4.7.3   DETAILED QUESTIONS AND SUGGESTIONS TO CONSIDER

1. Describe the power structure in your department college, and university. What is your place in this power structure?

2. Identify whether your department is cooperative or competitive. How are you functioning within this environment? Where can you go for advice on working effectively within either model or a combination of these models?

3. What are the most effective ways to communicate with your senior colleagues?

4. Your time is a commodity that you can use as leverage in a negotiation. Have you considered how you might make an exchange for taking on responsibility for new tasks with release time for such things as:

   - new course development?
   - advising load?
   - grant management?
   - administrative duties?

5. How do you delegate tasks? Are you comfortable delegating? When is it appropriate? To whom might you consider delegating such things as:

   - travel arrangements?
   - reimbursement paperwork?
   - purchasing?
   - budgeting?
   - meeting arrangements?
   - grading?
   - lab preparation?
   - lecture demonstration preparation?
   - course packet compilation?
   - photocopying?

6. Are there certain activities that you should avoid discussing with your colleagues? For instance, sometimes women faculty are labeled as "not-so-serious-researchers" if they express too much interest in teaching or outreach activities. Can you continue to do these activities at a level that it personally satisfying while maintaining the image that your colleagues expect?

BIBLIOGRAPHY

DEVELOPING NETWORKS, RELATIONSHIPS, AND MENTORING ACTIVITIES

AWIS, **A Hand Up: Women Mentoring Women in Science**, Association for Women in Science, Washington, D.C., 1995.

R. Boice, **Advice for New Faculty Members: Nihil Nimus**, Allyn and Bacon, Boston, 2000.

D.J. Dean, **Getting the Most out of Your Mentoring Relationships: A Handbook for Women in STEM,** Springer Science+Business Media, New York, NY, 2009.

R.M. Diamond, **Preparing for Promotion, Tenure, and Annual Review: A Faculty Guide**, 2nd Edition, Anker Publishing Com., Bolton, Massachusetts, 2004.

J.A. Goldsmith, J. Komlos, and P. Schine Gold, **The Chicago Guide to Your Academic Career: A Portable Mentor for Scholars from Graduate School Through Tenure**, The University of Chicago Press, Chicago, 2001.

R. Hayes, D. Grossman, **A Scientist's Guide to Talking with the Media: Practical Advice from the Union of Concerned Scientists**, Rutgers University Press, New Brunswick, NJ, 2006.

C. J. Lucas and J. W. Murray, Jr., **New Faculty: A Practical Guide for Academic Beginners**, Palgrave, New York, 2002.

S. Morgan and B. Whitener, **Speaking About Science: A Manual for Creating Clear Presentations**, Cambridge University Press, Cambridge, 2006. DOI: 10.1017/CBO9780511722066

A. C. Schoenfeld and R. Magnan, **Mentor in a Manual**, Magna Publications, 1994.

A.C. Schoenfeld and R. Magnan, **Mentor in a Manual: Climbing the Academic Ladder to Tenure**, 3rd Edition, Atwood Publishing, Madison, Wisconsin, 2004.

E. Toth, **Ms. Mentor's Impeccable Advice for Women in Academia**, University of Pennsylvania Press, 1997.

M. L. Whicker, J.J. Kronenfeld, R.A. Strickland, **Getting Tenure,** Survival Skills for Scholars Series, Vol. 8, Sage Publications, London, 1993.

Stress can do really strange things to people - both physically and mentally. I have seen my friends and peers exhibit a host of ailments and some of the oddest behavior during their untenured years - particularly, near the time of the tenure decision. The way to avoid having your job ruin your health and/or your personal relationships is to spend the time that you need taking care of yourself. Each individual is different, but it is likely that you know of at least one thing that will relax you without fail and one thing that you can do on a regular basis to keep healthy. The key is that you must take time out to do them.

CHAPTER 5

Getting Support and Evaluating Your Personal Health

For many assistant professors, accepting an academic position means moving to a new city and often to a different region of the country. When starting this new and challenging position, an enormous amount of time must be invested in the job. However, it is essential to also invest some time in life outside the university. For most, long-term happiness will depend on establishing new social networks, making friends, and becoming involved in the community. Pay attention to developing a balance between your professional and personal life early – right after moving to the new city – and don't put it off until feelings of loneliness and isolation set in.

Your faculty position will demand that you navigate, negotiate, and evaluate the boundaries between your personal and professional lives. The balance differs for each person, and only you can be the judge of what will constitute a healthy lifestyle for you. Many things will make the difference between thriving and merely surviving as an assistant professor.

5.1   YOUR LIFE OUTSIDE THE INSTITUTION

5.1.1   OVERARCHING QUESTIONS TO CONSIDER

- Have you made any personal connections in the community you live in?

- Can you identify community organizations or clubs that would interest you?

5.1.2   MENTORING CONVERSATION: ON THE IMPORTANCE OF A LIFE OUTSIDE OF WORK

It is easy to let your faculty position absorb every waking moment of every day. The truth is, I have let it do so for short periods of time when working on a specific project, but it is not good practice to allow that situation go on for the long term. It can be a recipe for a very lonely existence and won't give you the opportunity to develop support networks that you may need later.

A colleague of mine confided in me recently that he is worried he'll spend the rest of his life alone. He has tenure and a very successful career, but he focused so much on his work and so little on his social network that he is all alone in his nice big house. This is not what he had pictured for himself at this stage of his life, and he is not sure how to get from where he is to where he wants to be. He is a great guy, so I am hopeful that he can find the path.

As you develop in your faculty career, it is important to watch for signals concerning the balance between your personal life and your career. If your own mind and body are not getting your attention, then maybe an outside indicator will. For me, it was a huge red flag when my three-year-old told me "Mommy, don't bother me I'm writing my proposal." It broke my heart, but it also showed me what I already knew – I had been working way too much and letting things slide out of balance.

There are also professional benefits associated with having a well rounded life. I have noticed the correlation between having a life partner or outside interests and social intelligence. Leading an isolated existence and having a singular focus on one's work does not provide opportunities to practice social interactions. Let's face it, faculty members as a rule are sort of peculiar. We have to be, given that we are willing to spend so many years so engrossed in our specialty area. It can be easy to let our peculiarities rule and become difficult to deal with to the extreme of antisocial behavior. When we cultivate friendships and connections outside of academia it forces us to be a bit more socially savvy and capable of dealing effectively with everyday interpersonal interactions.

5.1.3   DETAILED QUESTIONS AND SUGGESTIONS TO CONSIDER

1. Have you allowed yourself time to find a place to live, shop, and enjoy recreational pursuits? Have you devoted enough emotional energy to adapt to your new surroundings?

2. Developing social networks both inside and outside of your institution is important. Have you set aside time and looked for avenues to develop personal friendships?

3. Have you met new people recently? Are you making friends and developing your personal network? Sometimes it is difficult to figure out how to meet people outside of your department, but you might consider the following avenues:

   • New faculty functions
   • Sports clubs
   • Volunteer organizations
   • Religious organizations
   • Parent organizations
   • Neighborhood association
   • Community events

4. It is possible to achieve your own personal balance, but the question is whether you do this privately or are vocal about it. Given the attitudes of the faculty in your department, is the best course of action to practice "discrimination avoidance"? For instance, do you want to consider:

   • minimizing the apparent intrusion of family life on academic commitments?
   • hiding one's care giving status?

• avoiding conversations that highlight gender or sexual orientation?

5.2   EMOTIONAL SUPPORT

5.2.1   OVERARCHING QUESTIONS TO CONSIDER

• In what ways will you develop the support structures you will need in the bad times?

• Do you know how to seek out help when you need it?

5.2.2   MENTORING CONVERSATION: ON GETTING THE SUPPORT YOU NEED

There are times in our lives when we will have to deal with new and challenging personal problems that are difficult to handle in isolation. Death of a close family member, illness of your child, substance abuse, disability, ...

I have been lucky so far in my life to have only had relatively minor challenges to face, but I found myself at my personal limits a few years ago when I was pregnant with my son. I turns out that I was not nearly as strong and enduring as I thought I was. Dealing with the nausea and fatigue I faced in the first trimester of my pregnancy day after day, week after week was mentally debilitating. I became despondent because of the relentless and grueling nature of the symptoms even though I knew eventually it would come to an end. After all, a pregnancy can't last much longer than 9 months.

What really helped was seeking out emotional support. We started a new mom's group in the college for faculty women who were pregnant or recently started a family. Although we officially met only once a month, those meetings and the email exchanges with the other new moms really helped me to focus on the light at the end of the tunnel. Eventually, there came a morning when I woke up and did not feel sick, and the next day I was not sick, and soon euphoria set in because the nausea was actually gone and the fatigue subsided. A few months later, I was rewarded with a beautiful baby boy! Of course, it turned out that he was not a good sleeper, so fatigue due to sleep deprivation was soon a problem, but I got used to it.

Although the first trimester of pregnancy can't compare to some of the other challenging and traumatic personal problems that others may encounter, the need for emotional support through these times is the same. When faced with such challenges, you must seek out the support you need. It may come in a variety of forms, but the keys are to seek out the support you need and to accept it when it is offered.

5.2.3   DETAILED QUESTIONS AND SUGGESTIONS TO CONSIDER

1. How much emotional support do you need? How much can you expect from others? Think about ways in which you can:

- communicate your feelings and thoughts to those close to you.

- optimize your schedule to set aside time for yourself and those close to you.

- find people with whom you can talk.

- make friends.

2. For women and men with children and/or other responsibilities outside the university, consider how much you can give to:

   - work,

   - your children, spouse, family, and pets,

   - your friends and social network,

   - the community, and

   - yourself.

3. Elder care can become an issue as we age. The challenges can be exacerbated for faculty who often do not live near their extended family. You may want to begin to prepare for these issues by asking the following:

   - Have you had a discussion with my parent(s) to ask about their plans as they age?

   - Can you identify community-based services for transportation, meals, home care, etc. that would assist with elder care? [1]

   - Is there a caregiver support group that would help you with emotional support and resource identification?

   - Is there an opportunity to move your aging parent(s) closer to your current location?

   - Are there arrangements that you can make with your schedule to allow flexibility for travel associated with elder care?

   - Does your institution have a stop-the-clock policy that includes time for illness of a family member?

5.3 YOUR PERSONAL HEALTH

5.3.1 OVERARCHING QUESTIONS TO CONSIDER

- How can you manage your stress?

- What techniques will you use to keep physically and mentally healthy?

---

[1] Department of Health & Human Services, "Eldercare Locator: Connecting You to Community Services," www.eldercare.gov (Accessed 11/11/09).

5.3.2   MENTORING CONVERSATION: ON THE PHYSICAL MANIFESTATIONS OF STRESS

I have observed a strong correlation between pretenure stress and the onset of physical ailments in the assistant professor population. Actually, it was first pointed out to me by my graduate advisor as he was approaching his own tenure decision. The internal and external pressures imposed on assistant professors are enormous. It is a highly demanding position for which we are only partially trained when we walk in the door. Even when good mentoring is available, the drive to succeed and the ever-increasing expectations impose substantial and often inhumane demands. In many, this stress takes its toll on the body and mind, which can present itself in real and tangible problems.

As an assistant professor, I saw the physical manifestation of the stress of pretenure life in myself and my junior colleagues. It is a plague on the junior faculty at many institutions, with symptoms such as severe weight gain, severe weight loss, stress fractures from excessive exercise, severe tinnitus, insomnia, and hypochondria. A friend of mine developed a facial tick, an unfortunate visible manifestation of the stress he was under. The facial tick served as a stress barometer and contorted his face fairly severely by the time the actual tenure decision was upon him. The day tenure was granted the tick disappeared, never to return. My own pretenure ailment, I feel, was the quintessential faculty ailment: one day I woke up and I was unable to shut my mouth! Unfortunately, it was not verboseness but a physical problem with my jaw brought on by years of clenching my teeth at night while I slept. If this is not a stress-related issue, I don't know what is.

Occasionally, remedies for fixing what is often considered to be a broken tenure system are proffered. More than once, I have heard it suggested that the tenure clock should be lengthened, and, indeed, many institutions have adopted "stop-the-clock" policies for child care and elder care. Unfortunately, I don't think these measures address the fact that the pretenure years are inherently stressful, and if we impose this excessively high level of stress on ourselves for an extended period of time, we begin to fall apart under the pressure. Until the system changes, you will have to do what you can to relieve the stress for yourself, to avoid the toll it will inevitably take on you if you don't.

5.3.3   DETAILED QUESTIONS AND SUGGESTIONS TO CONSIDER

1. Most people can't work 14 hours a day and stay mentally and physically healthy. Are you balancing your time so that you can be a whole and healthy person?

2. In what ways are you taking care of yourself by:

   - valuing yourself in a positive way?
   - standing up for yourself?
   - spending time with your family?
   - balancing for outside responsibilities?
   - meeting your own expectations, not those of others?
   - having fun?

   • keeping your job in perspective?

3. Make sure you are taking care of yourself physically with:

   • regular exercise,

   • enough sleep,

   • a healthy diet, and

   • regular checkups with your doctor and dentist.

4. How well are you using your time? How do you maintain equilibrium? How do you identify valuable and not-so-valuable uses of your time? It is not how much time you spend on
something that counts; what counts are results.

BIBLIOGRAPHY

GETTING SUPPORT AND EVALUATING YOUR PERSONAL HEALTH

S.R. Covey, A.R. Merrill, R.R. Merrill, **First Things First**, FranklinCovey Co., 1994.

W.H. Gmelch, **Coping with Faculty Stress**, Sage Publications, Newbury Park, CA, 1993.

I am frequently asked to give talks to graduate students and new faculty about career planning. I find this difficult to do. Although I think I am one of those people who is a planner, I rarely do it in a systematic way. Often, I make lists of my projects and future proposals, write down goals for long-term projects, and draw maps of different paths I might take. However, I rarely save these things and seldom look back at old versions. I think that is because I realize that my plans are constantly changing, and I don't want to be pinned down by what I thought I might do a year ago. Nonetheless, I think the exercise is an important thing for me to go through occasionally. It helps me focus on the immediate goals and plan my next move with a long term plan in mind.

Planning for the Future

The idea of mapping out the future at an early point in one's career scares many people. But if you realize that your plan will change and develop as you advance in your career, then you can think of it in a more creative and less permanent way.

Your strategic planning should take place on both a global and local scale. Reaching your overarching goals, like tenure, requires planning and coordinating a series of short-term goals, like publishing your next journal article or developing a new course. It is easy to get lost in the day to day obligations of faculty life and lose track of the big picture. You may have to work to make time for tasks associated with your global goals so that they are being pursued on a regular basis.

Strategic planning is a cyclical process that you build on continually throughout your career. You should revisit your plan on a regular basis and modify it to fit your new needs.

6.1 CHARTING YOUR FUTURE

6.1.1 OVERARCHING QUESTIONS TO CONSIDER

- What milestones can you set for yourself that will move you towards your goals for the future?

- What strategies work best for you when it comes to planning your work and prioritizing tasks?

6.1.2 MENTORING CONVERSATION: ON STRIVING FOR A MORE INTEGRATED SET OF LIFE GOALS

There is a notion, gaining in popularity, that an integrated approach to life goals is necessary for one's overall success and satisfaction. Proponents also suggest that having an unbalanced focus on one aspect of life, such as career, can ultimately have a negative impact on that very same aspect of life because of the chaos or dissatisfaction that occurs elsewhere. Stewart Friedman, for instance, advocates development as a leader through clarifying what is important, respecting the whole person, and developing creative solutions.[1]

It can be challenging to find these creative solutions in life where all of the stakeholders win and true integration is achieved with all of one's seemingly conflicting goals, responsibilities, and obligations. However, it is true that if you look at all of your life goals together, you will be able to develop a better coordinated plan for achieving what you want while minimizing clashes, both with yourself and those people around you who are important to your life.

[1] S.D. Friedman, **Total Leadership: Be a Better Leader, Have a Richer Life**, Harvard Business School publishing, Boston, MA, 2008.

One aspect of charting your future is timing. Setting your milestones relative to each other is a critical part of the equation. It may seem that the tenure mark is fixed in time, but that too can be moveable at many institutions with "stop the clock" policies. Even if the when of that particular milestone is fixed, the intermediate milestones need to be spaced out realistically, and milestones associated with other life goals need to be integrated within the overall framework. Whether these goals are starting a family, training for a marathon, or taking that trip to Europe, without some thought as to how these other goals fit in with the rest, there is a risk that they will get horribly delayed or the opportunity lost forever.

Creative synergies can be found. A colleague who wanted to spend time in Italy took Italian language classes, developed collaborations with colleagues at an Italian institution, sought out funding opportunities to perform his research abroad, and eventually ended up spending the better part of three years teaching and doing research in Italy. Clearly, this was many years in planning and preparation, but he achieved his personal goal and advanced his career at the same time.

Sometimes I even manage to make it all work out. As I write this section of the book, I am sitting in an apartment half way across the country from my home institution. My husband and I have always been challenged to juggle our respective faculty careers, but, occasionally, we can actually help to make opportunities for each other. Because we met after joining the university, our sabbatical schedules are out of phase. This year, he is on sabbatical and away from home for the semester. In many ways, it would have been simpler for me to have just stayed at home, but, instead, I arranged to do my teaching and committee work remotely (technology does have some advantages) so that the whole family could be together for a month in the middle of the semester. Although time consuming to arrange, there have been a number of benefits to my career on top of the obvious benefits to our son and marriage. I have been able to make new connections with several people on this other campus, be in close proximity with a company I am consulting for, make visits to alums for our university foundation, and have somewhat fewer interruptions allowing me some concentrated writing time for research papers and this book. The advance planning and integrated approach to the personal and professional has ultimately paid off.

6.1.3   DETAILED QUESTIONS AND SUGGESTIONS TO CONSIDER

1. One might think of dissatisfaction and satisfaction as opposites on the same continuum, but in the area of job satisfaction, it may not be that simple. Frederick Herzberg suggests that "The opposite of job dissatisfaction is not job satisfaction, but no job satisfaction."[2] In trying to achieve more satisfaction and less dissatisfaction in your job, you might focus on:

   - the negative aspects of your job that can be minimized or eliminated.

   - the positive aspects that can be made your central focus.

   - your personal goals and their alignment with the institution's goals.

[2] F. Herzberg, "One More Time: How Do You Motivate Employees," **Harvard Business Review**, 2002 (Sept/Oct 1987, 65(5), 109-120).

- creatively restructuring your job to meet your needs and goals.

2. What are your professional goals? How are they related to and influenced by your personal goals? Consider:

   - Maintaining a balance between your personal and professional life
   - Meeting your family obligations
   - Being in a geographic location that suits you
   - Choosing a research field that interests you and is a feasible pursuit in your current position
   - Moving between the public sector, industry and academe

3. Your goals provide a context for your decision making and the planning process. Some people purposely focus on achieving tenure as a goal while others avoid focusing on this particular milestone.

   - Have you considered how the goal of achieving tenure influences you?
   - Does it provide a focal point for you?
   - Does it create too much stress in your every day life?
   - What other global goals might you set that would satisfy you and put you on the path toward achieving tenure?

4. You should ask yourself several questions while developing your strategic plans:

   - What kind of time frame is appropriate for the goal I want to achieve?
   - What steps are involved in achieving this goal?
   - Who can help me identify the steps I must take and how to achieve this goal?
   - Who can support me and help me to achieve my goals?
   - What benchmarks can I use to judge my progress?
   - How can I stay on track once I have developed my plan?
   - How can I learn from what did not work in the past and deal with failure constructively?

5. There are also schematic methods that help some people to visualize their goals. Would charts, time lines, or graphs be valuable tools for you?

6. The average person changes careers every seven years. This is also true for many academics concerning their research program.

   - Are your skills broad enough to permit this flexibility?

- What kinds of continuing education will assist you as you navigate these changes?

- Are there opportunities to get support for a research shift in your field?

7. The demands of a faculty position make balancing family and career responsibilities difficult. Have you considered how your personal life fits into your career plans?

8. Many faculty plan to have children during the summer months, but this is not always possible![3] Many women faculty choose to put off having children until after tenure, but this carries risks. In making your plans to start a family, you might want to find out about the following:

- Maternity and parental leave policies

- Options for stopping the tenure clock

- Teaching relief the term after your child is born

- Precedents set by other faculty in the department or college

- Unspoken rules about pregnancy and maternity/paternity leave

- Attitudes of other faculty in your department towards family responsibilities

- The effect of children on the career success of other junior faculty

- University subsidized child care

- Arrangements for breast feeding

9. Are the goals you have set out for yourself with month/semester/year realistic? Are they in line with the expectations of others? What will happen if they are not met?

6.2 DOCUMENTING YOUR ACCOMPLISHMENTS

6.2.1 OVERARCHING QUESTIONS TO CONSIDER

- What strategy will work best for keeping track of all of your accomplishments over time?

6.2.2 MENTORING CONVERSATION: ON KEEPING IT ALL ORGANIZED

One straightforward thing that will make life much simpler when it comes time for your review or tenure promotion is meticulous record keeping. The simple act of regularly recording everything you have done that might be relevant to your case will save you enormous amounts of time in the long run. Don't worry about "what counts" or whether you are categorizing things right, just get it down in a file somewhere. For most, this would be easiest to do in a computer file, maybe even as a separate expanded version of your CV, but even a desk drawer where you throw papers, workshop fliers, and conference programs is better than nothing.

---

[3] R. Wilson, "Timing Is Everything: Academe's Annual Baby Boom," **The Chronicle of Higher Education**, June 25, 1999, http://chronicle.com/article/Timing-Is-Everything-Acade/2635/ (Accessed 11/11/09).

At some point, you will have to organize all of these things, certainly before your first annual or probationary review. This may be easiest to do if you follow the format that will eventually be required of your tenure packet. Seek out the rules stating what should be included and look at examples of recent tenure cases of your not-quite-so-junior-anymore colleagues. At many institutions, example cases are available through the office that handles tenure and promotion, but if they are not, then ask a colleague who has recently received tenure if you could look at his/her packet as an example. It is highly likely that you will not see anyone's complete packet – some pieces like the letters are confidential – but it will likely contain most of the content you will be responsible for providing. The structure of this document will provide a good framework for recording your accomplishments as you accumulate them.

Note that, particularly in the first or second year, it can be rather discouraging to look at someone else's successful tenure packet. Although they are only four to six years ahead of you, the gap between where you are and where you will need to be may seem terribly daunting. Don't let it create too much anxiety for you. You are building momentum in these early years and the example provides a target for the range you would like to be in. Although it may seem like an arms race, the tenure process should not have ever increasing expectations each year. I frequently talk with assistant professors who feel that they must do better than the last case put forward even if it was successful, or worry excessively about the rumors that X number of publications is not enough, and it must be at least Y. Try not to get sucked into the paranoia. Build yourself a strong case that you feel you proud of and that your department can support putting forward.

Until you are in the final editing stages of completing your packet, don't worry about what does and doesn't "count." Include it all in your documentation, you can categorize and edit out items later. I remember my first official mentor committee meeting vividly – at this institution, the mentor committee can have both a mentoring and evaluation role depending on how it is structured by the department. I had been keeping good record, put it all in the format required of the tenure packet, and came to the meeting proud of what I had been able to accomplish in the first two years. The committee members took their pens to it and crossed out all of the things that they felt would not "count." It seemed that there were only little scraps left by the end of the meeting. After the shock wore off, I realized that they were trying to help me identify areas in which I needed to strengthen. It was not that these items did not count at all, it was the level of importance placed on certain accomplishments. They wanted to be clear about what types of things would give me the highest probability of success. In the end, most of the items they crossed off were included in my tenure packet. Taken together, they built a well rounded case and showed a more complete picture of the type of faculty member that I had become.

6.2.3   DETAILED QUESTIONS AND SUGGESTIONS TO CONSIDER

1. Your curriculum vitae presents an image of you as a scholar or professional. Are you:

   - thinking about how best to document your experiences?
   - documenting your activities and accomplishments regularly?

- critiquing your CV periodically to assure balance and identify potential gaps?
- asking if the image presented in your CV fits with the type of person tenured by your department?
- getting feedback from senior faculty in your field?
- tooting your own horn?

2. Does your department keep your CV on file? Do you update it regularly? Try to add something to your CV every month. Doing this forces you to think about what you have accomplished, allows you to see the progress you are making, and paints a picture of how your career is developing.

3. Keep detailed records of your activities concerning:

- papers and conference proceedings,
- grants,
- seminars and invited talks,
- media coverage of your work,
- classes taught,
- student evaluations,
- committee work,
- students advised, and
- duties in the scholarly community.

4. Document the strength of your teaching by:

- seeking written peer reviews of your teaching by colleagues.
- retaining letters from students that offer praise for teaching, advising, or mentorship.
- obtaining copies of your student evaluations and departmental averages.

5. At several points, you will be asked to provide summaries of your teaching and your research. Do you have a draft of your teaching philosophy and your research statement that you revisit and update periodically?

6. Early in your career, you will face a formal review for tenure or reappointment of your contract. Your tenure packet or dossier is the mechanism used to present your case to your colleagues at these decision times. It is important to:

- determine what must be included in your tenure packet.
- find out who contributes to which parts of the tenure packet.

- begin collecting and organizing this information.

- obtain a copy of a recent successful tenure packet from a colleague.

- develop a strong case in all three areas of research, teaching, and service.

- review your tenure packet with your mentors and department chair periodically.

- address weaknesses in your case directly, documenting improvement in problem areas.

- understand all written policies and procedures regarding the review process.

- manage your case yourself by playing an active role in the process and the development of your dossier within the limitations of your institution.

7. Throughout your career, you will participate in performance reviews. Keep the following in mind:

- You retain control of your career regardless of whether reviews are positive or negative.

- Your department and institution has made an investment in you.

- The review is a two-way street.

- You must make your own priorities known.

- It is important to inform your department head and review committee about your past successes and future goals.

- You should solicit the assistance of others in achieving your goals.

- It is essential to be proactive, both in getting recognition for past achievements and in creating future opportunities.

8. Be prepared for a period of transition after your promotion. Have you talked to peers who are a few years ahead of you to find out about their post-tenure experiences? Have you considered what your next set of goals will be?

6.3   CAREERS OUTSIDE THE ACADEMY

6.3.1   OVERARCHING QUESTIONS TO CONSIDER

- What other career tracks would utilize the skills you have developed?

- How can you repackage your CV into a resume to effectively market yourself outside of academia?[4]

---

[4]M. Newhouse, "From CV to Resume," **The Chronicle of Higher Education**, Dec 3, 1999, http://chronicle.com/article/From-CV-to-R-sum-/45668/ (Accessed 11/11/09).

6.3.2   MENTORING CONVERSATION: ON LEAVING ACADEMIA

Faculty life can oftentimes seem like a long series of battles in an entrenched war. Knowing which battles to fight is a skill. It is not that you must only choose to fight the battles you think you will win, you must also consider the battles you will not forgive yourself for passing up. Not every battle is winnable and even the ones you do win will take a toll. They take time, energy, and often other resources that you could have used elsewhere. The costs can be very high and winning may not balance them out.

If you find yourself in a position where your job places you in a series of battles, or even forces you to choose between fighting and not fighting frequently, it may eventually wear you down. Some people thrive in this sort of situation, but many of us only find it exhausting and disheartening, particularly when the series of battles (even if they are little ones) seems to be unending. If facing these interminable battles comes with too high a price, you may decide that you need to give up the war.

A negative tenure decision may take you out of this war – and eventually you may even be thankful for the forced exit. I have talked to people who have told me years after being denied tenure that it was the best thing that could have happened to them. They have a happy and well adjusted life with a reasonable and enjoyable job out in the "real world." As James Jasper writes, "Contesting it is one way to regain your sense of control, although the downside is that you have to think about the whole mess for a long time, and the odds are stacked against you. At some point, and perhaps the sooner the better, you have to figure out what to do next."[5]

The important thing to also realize is that even after being granted tenure, you can still decide to leave the war if it is causing too much damage. I don't know anyone personally who has done so, but I hear stories whispered by my colleagues. I know a few who have made a partial departure by keeping the job but otherwise checking out. Some decide that the associate professor rank is the highest they need to attain. Others count the days to retirement and take it as soon as it is available, going on to start a second career. But few have the guts to just leave after they have fought and won that hard tenure battle. It would be a really hard thing to do, but in some cases, it may be the right choice. In the long run, the battle scars you will have to bear may not be worth it.

Whether you are forced out or you choose to leave, know that there is a big world outside academia where your skills and knowledge can be put to use. There are even jobs out there that pay more and have more reasonable expectations.

In some ways academia is a bit like a cult. Many of us faculty have never ventured outside the university walls since beginning as freshmen. We believe – because we have been told over and over again – that academia is the best place to be and that we have wonderful privileges here, like academic freedom and tenure, that we will find no where else. You know this, but I will say it out loud….. it is not perfect here in academia. There are many sacrifices for the privileges of academic freedom and tenure.

---

[5]J.M. Jasper, "Moving On After You Are Denied Tenure," **The Chronicle of Higher Education**, June 1, 2001, http://chronicle.com/article/Moving-On-After-You-Are-Den/45482/ (Accessed 11/11/09)

There are also those who stay in the academic world to teach because they see this as their calling. Some institutions will use your personal commitment to students to their advantage. In extreme cases, the institution's treatment of their faculty can borderline on abuse. If pursuing your calling is the only thing still keeping you in the war, know that you can pursue it in the outside world as well. There are many needy causes and organizations out there that may align nicely with what is important to you. They probably won't pay you more, but it is a safe bet that they will appreciate you more.

6.3.3   DETAILED QUESTIONS AND SUGGESTIONS TO CONSIDER

1. You may choose to leave your academic position at some point so you should always have a "Plan B" in mind:

   - Have you considered opportunities at other institutions, in industry, in the government sector, in a nonprofit organization, or as an independent consultant?

   - Are you interested in career paths not traditionally taken by people in your field?

   - Do you have other passions outside of your main field of study?

   - Identify different positions that might suit your interests and goals.

2. Keep your options open. In what ways do you explore new opportunities that you find challenging and exciting within your current position?

3. What if you don't get tenure? What will you do with this new beginning? Can you turn this seemingly negative outcome into an opportunity?

4. Can you take steps to demystify life after denial of tenure? Do you know someone who was denied tenure? Talk to them about the experience and their new career.

BIBLIOGRAPHY

PLANNING FOR THE FUTURE

R. G. Baldwin, "Faculty Career Development," **Current Issues in Higher Education**, No. 2. Washington, DC: ASHE-ERIC, 1979.

M. Morris Heiberger and J. Miller Vick, **The Academic Job Search Handbook**, University of Pennsylvania Press, 1996.

M. Newhouse, **Outside the Ivory Tower: A Guide for Academics Considering Alternative Careers**, Harvard University Office of Career Services, 1993.

J. H. Schuster, D. W. Wheeler, & Associates (Eds**.), Enhancing Faculty Careers: Strategies for Development and Renewal**. San Francisco: Jossey-Bass, 1990.

CHAPTER 7

Conclusion

At points in this job, your situation may devolve into just trying to survive. You may say to yourself like: "If I can just make it through this semester." "Get this proposal funded." "Have this publication accepted." But, for this job to be a career worth having, that you are happy with and fulfilled by, you must not just survive, you need to thrive. To make thriving possible, you need to understand the position fully. Identify the hard constraints. Find the soft constraints and see how much you can modify them. Turn the job into the right career for you. If that is not possible - if the fit just isn't right – then find something else to pursue for your career. Often, however, there is more flexibility than we had originally imagined and possibilities that are truly exciting. Explore these and make your faculty career one of your own design.

Index

Afterword

I was inspired to write this book by a wonderful self-assessment guide called "Thriving Through the Experience: An Assessment Guide For Graduate And Professional Students," which I co-authored with a group of women at the University of Minnesota. It was a project that I took on during my graduate studies, and it helped me to develop a broader scope for my own professional development at the time. The authors of that original manuscript, Joy Frestedt, Laboratory Medicine and Pathology, Wendy Crone, Aerospace Engineering and Mechanics, Katherine James, Veterinary Medicine, and Jessica Morgan, Anthropology, were all members of the Coalition of Women Graduate Students on the Twin Cities Campus of the University of Minnesota.

When I started as an Assistant Professor in the Department of Engineering Physics at the University of Wisconsin–Madison, I decided that I needed to build a new guide for myself that addressed issues faced by new faculty members. The development of this book, "Survive and Thrive: A Guide for Untenured Faculty" began in 1998. The book evolved over the years and was informed by the knowledge I have gained and my own personal experiences. Happily, I have received wonderful mentoring and advice from interactions with numerous individuals and through various workshops for faculty. I have also benefited from wonderful colleagues and generous peer mentors. Several people have given me feedback on the guide at various stages of its development, and I wholeheartedly thank them all. Special thanks go to Lindsey Stoddard Cameron and Prof. Laura McClure at the University of Wisconsin–Madison, particularly for their assistance in helping me to broaden the guide to address issues faced by new faculty across the disciplines.

Later in its development, the project was supported by the University of Wisconsin's Women Faculty Mentoring Program and the Women in Science and Engineering Leadership Institute (WISELI). This project was partially funded by a grant from the National Science Foundation (#0123666). Any opinions, findings, and conclusions or recommendations expressed in this material are those of the author and do not necessarily reflect the views of the National Science Foundation.

Your comments and suggestions are welcome and should be addressed to

Wendy Crone, Professor,
Department of Engineering Physics,
University of Wisconsin–Madison,
1500 Engineering Drive, Madison, WI 53706

or

crone@engr.wisc.edu

Author's Biography

WENDY C. CRONE

Professor Wendy C. Crone's research is in the area of solid mechanics, and many of the topics she has investigated are connected with nanotechnology and biotechnology. As a Professor in the Department of Engineering Physics with affiliate appointments in the Departments of Biomedical Engineering and Materials Science and Engineering at the University of Wisconsin-Madison, she has applied her technical expertise to improving fundamental understanding of mechanical response of materials, enhancing material behavior through surface modification and nanostructuring, exploring the interplay between cells and the mechanics of their surroundings, and developing new material applications and medical devices. She has worked in the medical device industry and has publications and patents pending on medical devices and biomaterials.

Professor Crone was granted a Faculty Early Career Award by the National Science Foundation, co-founded the MEMS and Nanotechnology Technical Division of the Society for Experimental Mechanics, and recently received the top Hot Talk/Cool Paper Award from the Materials Research Society.

Professor Crone is the current Director of the Women Faculty Mentoring Program at the University of Wisconsin-Madison. For two decades, the University of Wisconsin–Madison's successful Women Faculty Mentoring Program has served as a model to other colleges and universities across the country. She has also served as faculty Co-Director of the Women in Science and Women in Science and Engineering Residential Program and as the Director of Education for the Materials Science Research and Engineering Center on Nanostructured Materials and Interfaces.

65777839R00075